THE **ART OF** THE **LIE**

THE **ART OF** THE LIE

HOW THE **MANIPULATION OF LANGUAGE** AFFECTS OUR MINDS

MARCEL DANESI

Guilford, Connecticut

An imprint of The Rowman & Littlefield Publishing Group, Inc.
4501 Forbes Blvd., Ste. 200
Lanham, MD 20706
www.rowman.com

Distributed by NATIONAL BOOK NETWORK

British Library Cataloguing in Publication Information available

Library of Congress Cataloging-in-Publication Data

Names: Danesi, Marcel, 1946– author.
Title: The art of the lie : how the manipulation of language affects our mind / by Marcel Danesi.
Description: Lanham : Rowman & Littlefield, 2019. Includes bibliographical references and index.
Identifiers: LCCN 2019014272 (print) | LCCN 2019981002 (ebook) | ISBN 9781633885967 (paper) | ISBN 9781633885974 (epub)
Subjects: Trump, Donald, 1946—Language. | Truthfulness and falsehood—Political aspects—United States. | Rhetoric—Political aspects—United States. | Rhetoric and psychology. United States—Politics and government—2017–
Classification: LCC E912 .D36 2019 (print) | LCC E912 (ebook)
LC record available at https://lccn.loc.gov/2019014272
LC ebook record available at https://lccn.loc.gov/2019981002

∞™ The paper used in this publication meets the minimum requirements of American National Standard for Information Sciences—Permanence of Paper for Printed Library Materials, ANSI/NISO Z39.48-1992

CONTENTS

PREFACE

Half the truth is often a great lie.
—Benjamin Franklin (1706–1790)

In the 1972–1973 academic year, I got my first job as a professor of linguistics, at Rutgers University in New Brunswick, New Jersey. It was the period of the Watergate Hearings in Washington, which mesmerized my students and myself, becoming a major topic of discussion in two of my courses—a general linguistics course and one on Niccolò Machiavelli's *The Prince*. The focus of those discussions was on how Nixon's skillful deployment of language was a modern-day manifestation of Machiavellian cleverness and on how he was able to strategize his words to manipulate people's minds, especially his followers. In both classes, the linguistic features of lying were examined, analyzed, and discussed in the context of how Nixon's falsehoods were destroying the moral and ethical fabric of American society at the time. The conclusion we reached was that lies that come from the top echelons of any powerful institution, such as the White House, invariably lead to the unraveling of the principles of truth and justice that hold a democratic society together.

Skillful liars are dangerous people. They have the ability to twist words into weapons that can divide people against each other. Their words are designed to frighten and unnerve people, spurring them on to act in the interests of the liar, even if this goes against their own self-interest. Mendacious and deceitful language that is used strategically by those in power, such as the president of the United States, affects the mental health of everyone negatively, especially if it is repeated over and over like a nefarious, ritualistic chant. Nixon's lawyer during Watergate, John Dean, is recorded as saying to Nixon on one of the infamous secret tapes, "We have a cancer within, close to the presidency, that's growing."[1] The

Watergate Hearings were therapeutic, excising the emotional cancer that was spreading throughout America, as Dean so aptly put it. When the danger passed, after Nixon resigned in 1974, the cancer had apparently not metastasized and America regained its moral and emotional health, at least for a while.

The election of Donald J. Trump in 2016 brought back the cancer that had gone into remission, to belabor the metaphor somewhat. It manifested the same pattern of symptoms provoked by the same deployment of lies, deceit, denial, dissimulation, distraction, and duplicity. I had a foreboding sense of déjà vu, since I was again teaching the course in linguistics at the University of Toronto and also using Machiavelli in the class to bring out how language can be distorted to manipulate belief. Unlike Nixon, who was a statesman, Trump emerged in the political arena as a showman, a businessman-actor attracting many followers with the same kind of verbal bluster of classic American hucksterism, emblemized by circus impresario P. T. Barnum. I did not express myself in print during Watergate, perhaps because I was a rookie professor who had not quite yet gained the confidence to put on paper his thoughts about the relation of mendacious discourse from a Machiavellian prince to the destruction of social harmony. I have now gained that confidence, nearly half a century later, and feel impelled to share my linguistic expertise with an audience beyond that of my students.

In his 1987 book, *The Art of the Deal*, Donald Trump laid out a set of principles on how to negotiate business deals successfully by promoting oneself shamelessly via artifice and deception, never showing weakness or admitting fault. Recalling both Machiavelli's *Prince*, and the hyperbolic discourse of P. T. Barnum, Trump's book is a manifesto on how to lie, cheat, and confabulate in order to manipulate people's minds. During the early stages of the 2016 electoral campaign, Trump gave a televised interview in the lobby of Trump Tower on Fifth Avenue, in which he stated, "We need a leader that wrote *The Art of the Deal*."[2] To me, it was the first sign that the Nixonian cancer had suffered a relapse, since it was a lie—it was actually writer Tony Schwartz who had ghostwritten it. Significantly, Schwartz issued the following ironic tweet shortly thereafter: "Many thanks Donald Trump for suggesting I run for President, based on the fact that I wrote *The Art of the Deal*."[3] Schwartz realized that he had turned a pathological liar loose on society. Asked later about the book, Schwartz answered that he would have titled it *The Sociopath*.[4]

The *Art of the Deal* is a manifesto on how to promote oneself and pull off a deal by concealing or misrepresenting truth—a style of discourse that was actually given a name in the book—"truthful hyperbole."[5] The book could easily be retitled *The Art of the Lie*. I should state from the outset that my book is not only about Trump, but about the harmful linguistic "art" that he uses so skillfully and that is spreading broadly throughout the globe, becoming virtually unnoticeable in an age where mendacity seems to be nothing more than a communicative option. The internet is a "Brave New World," to employ the famous title of Aldous Huxley's 1932 novel—a novel that foreshadowed a society where psychological manipulation and classical conditioning were so common and widespread that they went unnoticed and were accepted as "normal." Trump's art has become an unconscious global vernacular—a lingua franca that spreads propaganda, conspiracies, misinformation, and disinformation on a daily basis.

Machiavellian mendacity aims to erode trust among people, while promoting its own self-interest. It incites acts of hatred and anger that are perpetrated in the world today, suggesting that human beliefs and fears can easily be manipulated with a simple twisting of words by masters of deceit. The art of the lie was described in detail for the first time in history by Machiavelli in his 1532 book, *The Prince*. In it, the Italian Renaissance philosopher advised rulers how to acquire and maintain power by both ethical and unethical methods. The successful ruler must be a "fox," able to baffle and deceive both his followers and opponents, while at the same time appearing as a "lion," feigning bravery and strength.

As president, Trump has shown himself to be a paragon of the Machiavellian liar-prince. He is a fox who knows how to outwit his opponents with mendacious ferocity, and he is a lion to his fans and followers, appearing strong and forceful. The question becomes, Why is an unscrupulous businessman, who has become a politician by happenstance, and who is an obvious liar, supported by so many people willing to believe him? Why is the language that he uses, which is clearly twisted, manipulative, and deceptive, accepted at face value? Do political expediency and ideological agendas override truth and objectivity at any cost? The purpose of this book is to investigate these questions, by looking at what lies do to people's minds, gleaning from the Trump phenomenon any relevant implications for the future of discourse and politics. Why do so many believe his lies? The answer may actually be a simple one, as encapsulated in the following statement made by the character George Costanza on the 1990s sitcom *Seinfeld*: "It's not a lie, if you believe it."[6]

I have written this book in the present tense, since I started drafting the manuscript right after Trump came to power. But it could be read at some future date in the past tense, since the argument that I will attempt to make—that lying is the most destructive of all types of deportment—will still be a valid one. Trump's sinister deployment of the Machiavellian Art of the Lie can then be viewed through the sober and dispassionate lens of retrospection. To use John Dean's phraseology again, my aim is to diagnose the "cancer" that mendacity "from the top" inevitably brings about, so that its "symptoms" can be exposed openly whereby, hopefully, they can be neutralized.

ACKNOWLEDGMENTS AND DEDICATION

I wish to thank my literary agent, Grace Freedson; my acquisitions editor at Prometheus Books, Steven L. Mitchell; and the many students who have encouraged me to write this book and who have given me so many insights and ideas into the discourse of lying over the years. I am especially grateful to Stacy Costa and Delaney Anderson, who read over an early draft of this book. Needless to say, any infelicities that it may contain are my sole responsibility.

I dedicate this book to my late uncle, Giovanni Bartolini, with whom I had numerous fruitful discussions about Nixon during Watergate and about the parallels between Nixon and Mussolini—under whose reign he had lived as a young person in Italy. I know that we would engage today in similar discussions about Trump, who also displays many similarities to Mussolini. I also want to dedicate this book to all those who have been victimized by lies and by falsehoods used against them. They are victims, not suckers. Once a lie has been perpetrated against someone, it is virtually impossible to eliminate its deleterious effects or to heal the emotional wounds that it inflicts. It is truly a cancer, as John Dean so aptly put it. And it must be exposed for what it is—a dangerous weapon of language.

1

LYING AS ART

We all know that Art is not truth. Art is a lie that makes us realize truth at least the truth that is given us to understand. The artist must know the manner whereby to convince others of the truthfulness of his lies.

—Pablo Picasso (1881–1973)

PROLOGUE

One of the most famous personages of ancient literature was renowned above all else for his clever lies, cunning, and resourcefulness. That personage was Odysseus, the king of Ithaca and the central figure of the Homeric epic *The Odyssey*. Centuries later, the Roman poet Virgil would credit Odysseus with the crafty ruse of the Trojan Horse, the hollow wooden statue of a horse in which the Greeks concealed themselves in order to clandestinely enter and invade the city of Troy. When Odysseus returned to his own kingdom after ten years of wandering, he continued to lie habitually, as if driven to do so by some inner compulsion, deceiving anyone who came into his sphere, including his wife, Penelope. For Odysseus, to speak meant, ipso facto, to lie.

A plausible subtext of Odysseus's story is that lying may well be an intrinsic part of human nature, not a deviation from it. However, the ways in which the hero Odysseus perpetrates his lies go well beyond how and why ordinary people tell them. The late classicist Peter Walcot refers to Odysseus's mendacity as an "art," calling for a particular kind of ability or adeptness at manipulating the meaning nuances of words and the flow of conversations.[1] This may be why Odysseus stands out from common folk in Homer's poem. He has a special talent that he can use at will to insulate himself from verbal counterattacks and other forms of opposition against him. Homer was obviously intrigued by the persuasive power of artful mendacity, describing Odysseus with adjectives such as the "many-sided Odysseus," "resourceful Odysseus," "devious Odysseus," and "subtle Odysseus." To this day, we take pleasure in reading about Odysseus's exploits, admiring his art of the lie as a manifestation of uncommon intelligence. We do not see his lies as strictly immoral, but as strategies for gaining success. Odysseus uses deception, for instance, to defeat the cyclops, who is much bigger and stronger than he is and thus a greater danger to the world.

The stories of legendary liars fascinate us to this day. Lying is a theme found in the folklore of all peoples. Everyone lies. We lie to avoid negative repercussions, to evade trouble, to circumvent hurtful facts, or to protect our self-image. No one has ever taught us to lie. It emerges spontaneously during infancy, revealing an unconscious verbal know-how that we use instinctively to gain some advantage over someone or to avoid adversity by twisting the meanings of words. Analogues to human lying exist in other species. A chimp foraging for food will often pretend not to have noticed a food source to avoid alerting other chimps about its location. The chimp will then hide somewhere and pounce on the food when no other chimp is around. This shows an instinctive ability to discern another chimp's intentions and to act purposefully on it.[2] But there is no real equivalence between such forms of animal deception and human lying, since the latter requires language. As journalist Robert Wright observes in *The Moral Animal*, human deception goes well beyond instinctive or reactive behavior, involving the conscious ability to manipulate someone else's mind with words.[3]

Throughout history, distinguishing between truth and lies has been a central objective of philosophy and theology. The fall of humanity from paradise, as recounted in the Bible, ultimately comes from a temptation perpetrated by the first liar of the heavens, Lucifer, who deceived not only Adam and Eve but also the other angels with his duplicity and deviousness. In John 8:44 he is described as "a murderer from the beginning, . . . a liar, and the father of it."[4] Lucifer knew that humans are vulnerable to lies, which he himself used nefariously to control their minds so that they would do his bidding.

Similar stories of human origins are found across cultures; in them, the conscious use of lying is typically seen as an act of free will. These are cautionary tales about the power of lies to control minds and alter human destiny. As Prometheus stated in Aeschylus's great ancient drama, *Prometheus Bound*, the capacity for lying has ensured that "rulers would conquer and control not by strength, nor by violence, but by cunning."[5] One of the first manifestos on political and military warfare, written around 500 BCE by Chinese military strategist and philosopher Sun Tzu and called the *Art of War*, identifies a set of principles on which war is based. In it, Sun Tzu suggests (like Prometheus) that the most effective and consequential victories are those that are gained through artful deception. As he insightfully put it, "All warfare is based on deception. Hence, when able to attack, we must seem unable; when using our forces, we must seem inactive; when we are near, we must make the enemy believe we are far away; when far away, we must make him believe we are near."[6]

It can be argued that the motivation for the creation of ethical and moral codes across the world and across time has been to counteract the deleterious effects of lying. Aristotle held that virtues such as justice, charity, and generosity benefitted both the person possessing them and society generally, implying that these are necessary antidotes to the destructive effects of lies and deceit.[7] The eighteenth-century philosopher Immanuel Kant saw truth, honesty, and integrity as central to ethical behavior, advising people to respect each other in order to preserve the moral order.[8] Since antiquity, we have held up ethical behavior as our main protection against mendacity and deceit—as the only way to counteract Lucifer's original act of lying. Lucifer is described, appropriately, as the "prince of lies." The analogous term "liar-prince" will be used throughout this book to refer to the masterful liar who has the same kind of talent for using language to deceive and manipulate people's minds.

Lying manifests itself in a host of verbal behaviors, from simple fibbing to sinister dissimulation. Understanding what lying is, in an age of political mayhem and social media falsehoods, has become urgent since, as Prometheus warned, deception is the most nefarious of all political strategies. The liar-prince can fabricate falsehoods on the spot for opportunistic reasons, veiling them ingeniously as truths. As a consequence, he can exploit beliefs advantageously. He is skilled, in short, at the "Art of the Lie," using it to affect the course of events, for better or for worse. The purpose of this opening chapter is to set the stage for decoding the features of this unethical "Art" in subsequent chapters.

LIES AND LYING

Colloquially, we pigeonhole lies into two broad categories—"white" and "black." The former are trivial and perceived to be largely harmless, told normally to avoid offending or hurting someone's feelings or else to sidestep embarrassment or potential imbroglios. If a friend asks us whether we remembered to mail something and we answer yes, even though we have not done it yet (but will), we are telling a white lie. White lies make it easy for us to avoid appearing in a bad light or to dodge reprobation, such as spinning a simple tale to "explain" why we got home late. Although white lies may be innocuous, eventually they will have a cumulative detrimental effect on people's interpersonal relations. The "black" lie has, literally, a "darker" function than the white lie. It is designed to negatively affect others, not just avoid an uncomfortable situation. It is this use of lying that falls under the rubric of the Machiavellian Art of the Lie. It is little wonder that Lucifer was called the "Prince of Darkness" by poet John Milton in his 1652 epic poem *Paradise Lost*. The same designation is used in Manichaeism to refer to the force of darkness that undergirds human destiny.

As mentioned, no one has ever taught us to lie. We do it instinctively from the moment we start realizing the power of language to regulate and influence the opinions and reactions of those around us. Some psychologists see the emergence of lying in childhood as a developmental milestone, calling it, rather sardonically, the stage of *Machiavellian intelligence*,[9] defined as the ability to project oneself into the minds of others so as to manipulate them for self-advantage. As psychologist Richard Byrne explains:[10]

> The essence of the Machiavellian intelligence hypothesis is that intelligence evolved in social circumstances. The individuals would be favoured who were able to use and exploit others in their social group, without causing the disruption and potential group fission liable to result from naked aggression. Their manipulations might as easily involve co-operation as conflict, sharing as hoarding—but in each case the end is exploitative and selfish. . . . Consistent with the Machiavellian intelligence hypothesis, social species of primates display both complexity of social manipulation and considerable knowledge of social information. This social complexity needs to be fully appreciated, to understand the strength of the case for Machiavellian intelligence.[11]

The skilled use of black lies for self-advantage will be called Machiavellian throughout this book, defined as the talent for selecting and assembling words to produce falsehoods that will normally escape detection, like a magical illusion trick. The Machiavellian liar is a master illusionist, who performs verbal wizardry to intentionally deceive people. The Art of the Lie is his textbook. (I should mention that I use masculine pronouns in reference to the Machiavellian liar throughout, because the liars to be discussed in this book are all male.) University of Louisville linguist Frank Nuessel also characterizes the black art of lying as an art of illusion, based on a special innate ability to weave deception and dissimulation into common discourse.[12]

Lying is common, manifesting itself in a variety of deceptive strategies described by English vocabulary with words such as *swindle*, *defraud*, *cheat*, *trick*, *hoodwink*, *dupe*, *mislead*, *delude*, *outwit*, *lead on*, *inveigle*, *beguile*, *double-cross*, among others. A rapid anecdotal probe of three other languages—Italian, French, and Russian—reveals a comparable listing of terms.[13] But lying may not be a universal trait. Some languages have significantly fewer words for lying, implying, perhaps, that some cultures may not have developed the so-called Machiavellian intelligence to the same degree, if at all, because of their different historical experiences and traditions.

Con artists, hucksters, and duplicitous people are "natural born liars," possessing the ability to easily fool unsuspecting people anytime and anywhere. Their skillful use of deceptive language to persuade or dissuade others has a fiendish aim—to dupe people into silence or compliance, as the case may be. They can also instill fear in those who see through them because people know intuitively that the masterful liar can utilize his skills against them, destroying reputations and friendships in the process. Manipulation and fearmongering are primary goals of the liar-prince. These allow him to rise to leadership by forging alliances, gathering followers and allies, and offsetting opponents through his mendacious art. The allies typically come under his direct mind control; the followers see in him a lion warrior; the opponents fear that he will insidiously destroy them publicly with his words. Writers have dealt with the power of mendacity from the time of Homer to the present day. Characters such as Shakespeare's Iago and F. Scott Fitzgerald's Jay Gatsby are scary because they have the ability to control others with their lies, evoking fear not with physical prowess but with wit and cunning. One of the most emblematic of all the great liars in literature is Shakespeare's Falstaff, a dominant figure in *Henry IV, Part I*; *Henry IV, Part II*; *The Merry Wives of Windsor*; and *Henry V*. Falstaff is a self-indulgent liar, coward, and braggart. He spends a large part of his time at an inn, where he presides over a group of rascals and scoundrels who are attracted to him. If life imitates art, then one could plausibly characterize someone like Donald Trump as a real-life version of the Falstaff character—a consummate liar who knows how to lie to attract people into his realm of influence through cunning. Like Falstaff, Trump also has a comedic charm about him that is self-serving and attractive to his followers.

The focus in this book is on the art of the liar-prince and his strategic utilization of duplicity, deceit, subterfuge, and confabulation to achieve and maintain political power. He knows how to sow division with words and affect the course of events through them. Liar-princes have abounded throughout human history. The Dreyfus Affair is but one example. French Jewish army officer Alfred Dreyfus (1859–1935) was falsely accused in 1894 of selling military secrets to the Germans. His trial and imprisonment caused a major political crisis in France. Anti-Semitic groups used the falsehood to stir up racial hatred. As it turned out, the incriminating evidence was forged by an army major, Charles Esterhazy. It was an example of what today we would call "fake news." This kind of event has occurred throughout history and across societies. Lies such as the one by Esterhazy are particularly destructive because they tap into prejudices that may be buried unconsciously, stoking feelings of resentment against a targeted group. A quote commonly attributed to either Adolf Hitler or his minister of propaganda, George Goebbels, encapsulates the foregoing discussion perfectly: "Make the lie big, make it simple, keep saying it, and eventually they will believe it."[14] Portraying Nazism and Stalinism as having a source in prejudice, political theorist Hannah Arendt observed that the reason why conspiracies and lies (such as the Dreyfus Affair) have such drastic effects is "not that you believe the lies, but rather that nobody believes anything any longer."[15]

The liar-prince thwarts the existing social order with his ability to make his lies seem truthful and believable. As Socrates perceptively noted, "Whenever, therefore, people are deceived and form opinions wide of the truth, it is clear that the error has slid into their minds through the medium of certain resemblances to that truth."[16] Fascinated by lies and how they have the power to damage the human spirit, the early Christian theologian St. Augustine of Hippo (354–430 CE), wrote two treatises on lying—*De mendacio* ("About Lying") and *Contra mendacium* ("Against Lying"). He argued that all lies are unethical, no matter how harmless their effects might be (as in the case of white lies), because we are all susceptible to falsity and deception. As the Dutch humanist, Desiderius Erasmus, perspicaciously observed, "Man's mind is so formed that it is far more susceptible to falsehood than to truth."[17]

Perhaps at no other time in human history has mendacity found such a fertile ground for stoking prejudices and hatred as in the contemporary world of social media, where conspiracy theories and fake news are so common that they go largely unnoticed as such. In this intellectually amorphous environment, truth and lies, facts and untruths, myths and science compete for people's minds. It is an environment, as will be argued in this book, that has empowered petty liars to gain fame. In this electronic mind fog, as it can be called, liar-princes of all political stripes emerge as heroes, gaining prominence through the "chatter" that occurs throughout the fog.

A MACHIAVELLIAN ART

The adjective *Machiavellian* is used in English (and other languages) in reference to ruthless liars, deceivers, scammers, and swindlers. The Renaissance Italian statesman and political philosopher Niccolò Machiavelli (1469–1527) saw lying as the most effective game plan to acquire and maintain political power. In chapter 18 of his manifesto, *The Prince*, Machiavelli laid out a psychological and political blueprint for manipulating people's minds.[18] Through intentional mendacity the liar-prince will attract followers and allies and be able to forge alliances, not by force, but by well-chosen words. To gain the upper hand, the liar-prince must fashion his words to stoke anger or antipathy against the status quo. This motivates those who feel disenchanted or resentful to rise up and defend him, shielding him from counterattacks and willing to do anything to help him maintain power. The prince's aim must be to create a sense of purpose, whether real or imaginary, among the followers. The liar-prince and his acolytes are bonded by an unconscious "all for one, one for all" worldview.

Machiavelli's book is unique since (as far as I can tell) no similar manifesto had existed prior to it. It went contrary to all philosophical and religious traditions, which have always portrayed lying as one of the most destructive (and sinful) of all human behaviors. Even when the reason for lying is to protect oneself, it is never ethical to lie, since it destroys morals and virtues. On the other side, in his controversial yet penetrating 1878 treatise, *Human, All Too Human*, Friedrich Nietzsche saw truth-telling as a weakness. Like Machiavelli, he saw mendacity cynically as the fuel propelling social progress.[19]

From the outset, Machiavelli makes it clear that lying is the most effective political-military weapon because it can influence minds, circumventing reason. He puts it as follows:

> Everyone admits how praiseworthy it is in a prince to keep faith, and to live with integrity and not with craft. Nevertheless our experience has been that those princes who have done great things have held good faith of little account, and have known how to circumvent the intellect of men by craft, and in the end have overcome those who have relied on their word.[20]

The experience Machiavelli had gained as a government official, and his study of the history of Florence, led him to view politics as fundamentally corrupt. Previous philosophers had treated politics idealistically, within the framework of ethical and moral behavior. But Machiavelli sought to explain the nature of politics realistically, at least as he saw it.

With few exceptions, he saw humans as naturally inclined to be dishonest (for self-gain). The emphasis on ideals was thus illusory, no matter how much we strived to pursue them. Machiavelli portrayed the state figuratively, as an organism with the prince as the "head" of the "body." Extending the metaphor, he described a "healthy" state as orderly and in balance, allowing its denizens to experience happiness and security. An "unhealthy" state requires strong measures to restore its health. The liar-prince must know how to dupe people into believing that he, and he alone, can restore the state to health. To do this, he cannot be bound by traditional ethical norms. He should be concerned only with strategies and actions that will lead to his own success. It is ironic to note that Machiavelli may not have taken his own advice. He had organized a political coup against the powerful Medici family, which ruled Florence. But in 1512, the attempt collapsed. The Medicis regained power, and Machiavelli was arrested, imprisoned, and tortured on suspicion of plotting against them.

A key strategy that the liar-prince must learn to deploy effectively is how to exaggerate or magnify discourses in order to befuddle or confuse people. In Trump's book, *The Art of the Deal*, the primary means for doing so is through what he calls "truthful hyperbole," a phrase that resonates with the circus culture of nineteenth-century America, when bombastic and hyperbolic speech was part of the show's allure. P. T. Barnum, the entrepreneur and circus impresario, may have been the first to employ this type of speech to promote his spectacles, calling his circus the "Greatest Show on Earth." A similar type of hyperbolic language is now a common ploy in advertising and in sales promotions—a theme that will be discussed in chapter 7.

Machiavelli maintained that being truthful is an Achilles heel that the liar-prince must avoid at all costs. Communication must be devised in such a way that followers and allies will not be aware of the deception. If it should be noticed, excuses must be readily concocted through denial and deflection. It is remarkable to read Machiavelli's principles of mendacity today, witnessing their implementation in Trump's tweets, statements, and speeches, as will be discussed later in this book.

Machiavelli suggests that, overall, the liar-prince must strive to be both a fox and a lion: "A prince, therefore, being compelled knowingly to adopt the beast, ought to choose the fox and the lion."[21] The fox is clever and can easily recognize traps, seeing through the counter-deceptions of others; the lion is the utmost figure of bravery and strength. The prince needs to be both, using cunning to fend off others and the performance of strength in front of his followers to maintain their loyalty. So, when someone accuses the prince of lying, the best strategy, as Machiavelli emphasized, is to be a fox and discover if there is any snare in it, and then assume the persona of a lion, using the same snare to put the accuser on the defensive by throwing it back at him. Machiavelli was not an ideologue or a moralist—he realized practically that a ruler had to adapt to circumstances—to be a fox or lion when required. The most crucial strategy in all this is that a prince should always employ dissimulation when the situation puts him at a disadvantage: "Therefore a wise lord cannot, nor ought he to, keep faith when such observance may be turned against him, and when the reasons that caused him to pledge it exist no longer."[22]

So, is there any effective defense against the Machiavellian liar-prince? Can a fox who acts like a lion ever be exposed and rendered ineffectual? The liar-prince eventually tires people out with his lies, which will arguably be a major factor leading to his demise. Moreover, as Machiavelli himself knew, the greatest danger for a masterful deceiver is to be outdeceived, and this happens more frequently than one might think. This topic will be discussed in the final chapter.

LANGUAGE, BELIEF, AND REALITY

Lying has always been a discourse tool of criminal organizations, such as the Mafia, which continues to use dissimulation and falsification to carry out its unlawful activities. The word *Mafia* was documented for the first time in an 1868 dictionary, where it is defined as "the actions, deeds, and words of someone who tries to act like a wise guy."[23] As sociologist Diego Gambetta points out, the term was a fiction, "loosely inspired by the real thing," that "can be said to have created the phenomenon."[24] It is, in other words, a classic case of *confabulation*, or the creation of a falsehood that becomes believable after the fact, gaining semantic sustainability over time. At the time many criminal gangs existed, but they were perceived essentially as groups of casual or random street thugs. The Sicilian name *mafiusu* was being bandied about to provide a useful designation for them. When it became a moniker, it pinpointed a particular group as a distinct organization, separating it from common thugs. The case of the Mafia reveals a fundamental principle of human cognition—there is no "reality" without a name for it. It is worthwhile repeating here what the anthropological linguist Edward Sapir wrote about this, since the link between language and reality is a critical one for dissecting and neutralizing the endemic threat posed by master liars:

> Human beings do not live in the objective world alone, nor alone in the world of social activity as ordinarily understood, but are very much at the mercy of the particular language which has become the medium of expression for their society. It is quite an illusion to imagine that one adjusts to reality essentially without the use of language and that language is merely an incidental means of solving specific problems of communication or reflection. The fact of the matter is that the "real world" is to a large extent unconsciously built up on the language habits of the group.[25]

A lie distorts that "real world," by manipulating words to create a false or misleading depiction of it. Criminal groups existed in Sicily long before the Mafia, but the coinage of the word "Mafia" in the late nineteenth century provided a collective label for them; without it, they would have been relegated to the social wayside as nondistinctive obscure hoodlums. It allowed gangsters to concoct an identity for themselves, as an "honor society." As Mafia historian Paul Lunde aptly remarks, the "lie of the Mafia as a historically based society has been a disastrous one for Sicily."[26] Already in 1900, Antonino Cutrera, an early anti-Mafia activist and a public security officer, wrote, "For historical and ethnographic reasons, Sicily has for many years suffered a social vice perpetrated on it by the Mafia. This vice has hindered its social development and has compromised the thrust of its civilization."[27] As Machiavelli certainly knew, lies affect the course of history by creating false beliefs that spread throughout a collectivity unsuspectingly. There would be no Mafia, as it has developed today, without the initial lie through which it emerged.

Falsehoods are never believed in the abstract; they must be devised as referring to something that people can understand concretely or to which they can relate personally. One of the slogans that Trump used throughout the campaign and into his presidency is "Drain the Swamp." This was designed to stoke the resentment that his followers harbored unconsciously about Washington politics, eliciting a graphic mental image of the previous government as mired in corruption. It also added fuel to the feeling of many that a "liberal elite" had taken over America, thwarting its traditional religious and blue-collar values. The "elite" is thus perceived as an "enemy" of America and the mainstream media as the accomplice, ignoring the values and views of "real" Americans, looking down on them as antiquated or ignorant. Whatever the "liberal media" say about Trump, he can now dismiss it as the "fake news" of those in the "swamp." By repeating such slogans and catchphrases over and over, the mental images they generate become entrenched in many people's minds suspending their ability to them as metaphors. It is a brilliant Machiavellian strategy, stoking unconscious resentments via metaphor. People at Trump's rallies love to hear this type of language, which generates a euphoric "high." Trump's base is thus prepared to stay with him to the end—no matter what the consequences are, including the risk of losing everything. To quote Sun Tzu, again, the clever subterfuges of the liar-prince cause "the people to be in complete accord with their ruler, so that they will follow him regardless of their lives, undismayed by any danger."[28]

Trump is a master Machiavellian liar. In tweets and speeches, he lies repeatedly about the presence of an "enemy of the people" within America, namely the liberal elite, who make up the swamp, which must be drained, so that the real America can retrieve its past greatness. Significantly, although it goes back to Roman times, the phrase *enemy of the people* was used repeatedly in Soviet Russia as a means to terrify people about a hidden, destructive antisocial force lurking within the society.[29] So, in order to "Make America Great Again" (MAGA), the enemy within must be drained out of the swamp so that the real America can reemerge.

Oscar Wilde once wrote that "Life imitates Art far more than Art imitates Life."[30] He was challenging the long-standing Aristotelian notion of mimesis, or the theory that art is an imitation of life. Wilde turned this notion on its head by asserting that "the self-conscious aim of Life is to find expression."[31] Wilde knew that reality and our representations of it are perceived unconsciously as one and the same. He used the example of the London fog to make his case. Although fog has always existed in London, one notices its qualities and effects because "poets and painters have taught the loveliness of such effects. They did not exist till Art had invented them."[32] To extend Wilde's view, Trump is a master at creating a mental "fog," with his consummate ability to create persuasive metaphors such as the "swamp" one. Most of Trump's followers live in the fog he has generated. They perceive his misleading language as a strategy in the overall battle to take back America. It is an example of avoiding unwanted outcomes by ignoring the lies of someone admired.[33] As Samuel Butler so insightfully put it, "Belief like any other moving body follows the path of least resistance."[34] Trump's lying is perceived by his base and allies as part of an ongoing cultural warfare, and thus a necessary tactic for battling the enemy within.

Belief is a paradoxical feature of the human mind, since it is often shaped by events that are false, but which people perceive as true nonetheless. This was brought out by a famous 1940 study, *The Invasion from Mars: A Study in the Psychology of Panic*, by Princeton psychologist Hadley Cantril and his research team. Their research project aimed to unravel the reasons for the panic created by the 1938 radio broadcast of a docudrama based on H. G. Wells's novel about interplanetary invasion, *War of the Worlds*.[35] Many listeners believed that the broadcast was real, despite periodic announcements during the broadcast that it was a fictional dramatization. Some residents in the New Jersey area (where the invasion was purported as occurring) fled their homes, calling the local authorities in a hysterical state of mind. After interviewing 135 subjects, Cantril concluded that the key factor in the panic was educational background—better-educated listeners were more likely to recognize the broadcast as fake than were less-educated ones, who were the ones most likely to believe that it was real and to react emotionally.

The study was criticized on the grounds that it did not establish a true correlation between the radio broadcast, the degree of reported panic, and the educational backgrounds of the listeners. Moreover, the panic may have been caused by subsequent media stories that intentionally exaggerated the panic. No deaths or serious injuries were ever linked to the broadcast, and the streets were never overly crowded with hysterical citizens running around in panic as the media claimed. The reported panic may have itself been a media exaggeration, with headlines such as, "Radio Fake Scares Nation" (*Chicago Herald Tribune*, October 31, 1938); "Radio Listeners in Panic" (*New York Times*, October 31, 1938); "Fake 'War' on Radio Spreads Panic over U.S." (*New York Daily News*, October 31, 1938). Nevertheless, the fact remains that some people did react hysterically, perhaps because they were already predisposed to do so by believing in space aliens in the first place—a factor that was not taken into account by the researchers. This topic will be discussed in more detail in the final chapter.

DISCOURSE

We hardly realize the influence that words have on us as we talk to each other. In the 1920s, Russian literary critic Mikhail Bakhtin saw discourse as directive of social behaviors and an intrinsic part of how we construct ideologies or worldviews, political and otherwise.[36] Trump has clearly grasped this principle of discourse. His metaphorical slogans—the "swamp," "enemy of the people," "fake news," and "MAGA"—are in fact a type of easily recognizable "Trumpian discourse," through which he forges images that tap into fears and ill feelings. For example, his attacks against environmental protection measures tap into the dread over the loss of jobs that would ensue from them:

> Our precious national treasures must be protected. And they, from now on, will be protected.[37]
>
> I've spoken with many state and local leaders—a number of them here today—who care very much about preserving our land and who are gravely concerned about this massive federal land grab.[38]
>
> California wildfires are being magnified & made so much worse by the bad environmental laws which aren't allowing massive amounts of readily available water to be properly utilized. It is being diverted into the Pacific Ocean. Must also tree clear to stop fire from spreading![39]

The phrase *our precious national treasures* reverberates at many levels of unconscious meaning among some people. At one level, it alludes to the sense that the federal government has overstepped the people's will by taking control of the environment. This purported subjugation of the people's will is anathema to the adherents of MAGA, including the radical conservatives in Congress who operate as an ideological faction under the aegis of "Freedom Caucus." This subtle message is reinforced in the second statement above, with the claim that people "are gravely concerned about this massive federal land grab" and, in states such as California, are subjugated by "bad environmental laws" (third statement). Trumpian discourse works this way—it amalgamates images cohesively to stoke fears of government controlling American lives by putting the environment above them with its foolish laws.

Trumpian discourse is an example of *bricolage*, a term used by anthropologist Claude Lévi-Strauss in his 1962 book, *La pensée sauvage*, to characterize the sense of magic that ritualistic language evokes in members of a group, binding them together indelibly.[40] To outsiders, the same words used over and over may appear to be nonsensical, but to insiders, they reinforce a belief system that keeps the group united against perceived enemies. Trump's most fervent followers at his rallies typically love to hear the same attacks, wisecracks, jokes, and disparagements of the "enemies" (liberals and intellectuals), making them feel united in the quest for a greater good (MAGA). Any breakaway from this type of discourse would literally break the magic spell, and might even lead to Trump's decline as a leader. As social scientist Wilson Bryan Key suggests, this type of discourse is effective because, like the rhetorical oratory at religious revival meetings, it is designed to stir up emotions and impart a sense of meaningfulness above and beyond the moment, projecting the ritual into the domain of the spiritual.[41] Trump's rallies are, in a phrase, bricolage performances that stir people's emotions, making them feel part of a moral quest, while entertaining them at the same time. As Oscar Wilde once wrote, "The liar at any rate recognizes that recreation, not instruction, is the aim of conversation, and is a far more civilized being than the blockhead who loudly expresses his disbelief in a story which is told simply for the amusement of the company."[42]

The MAGA slogan requires initial commentary here, even though it will be discussed throughout this book. On the surface, it seems simply to evoke an image of an idyllic past, free from the moral relativism of liberalism that many feel beset America under previous presidencies, especially that of Barack Obama. But below this surface, it evokes a phobia of otherness—of anyone who is not racially and culturally white. As a skilled Machiavellian deceiver, Trump realized from the outset of his campaign to become president that he could tap into this phobia with his "birther" claim that Obama was a Muslim who had not been born in the United States. His adoption of this egregious conspiracy theory had an instant and powerful impact on those who were dissatisfied with the style of liberal government that Obama and his followers espoused. The birther falsehood stoked resentment in a large group of people, making it virtually immune from counter-argumentation. Attempts to dispel the birther claim, in fact, have gone awry among Trump's followers, although Trump no longer mentioned it after he had gained the presidency, and begrudgingly admitted that it was not true. In effect, he no longer needed this falsehood in an overt way, since it became embedded unconsciously in the MAGA narrative, which envisions a social world that would, by implication, never have allowed an African American to become the president.

MAGA is a symbolic rallying cry in a cultural civil war. It recalls Italian Fascist dictator Benito Mussolini's (1883–1945) similar rallying cry for a recovery of Italy's imperial Roman past. Mussolini even used the Roman hand salute at his rallies to symbolize the importance of retrieving this past. Trump eerily employs an analogous type of body language to that of Mussolini at his rallies, raising his head imperiously after reciting a falsehood (in a way that is strikingly similar to Mussolini's head lifts during his speeches). Trump's opponents act as if his lies and his ludicrous postures at rallies will eventually bring him down, when in actual fact they have raised him up, as writer and political adviser Amanda Carpenter has so cogently argued in her important book, *Gaslighting America*.[43] Followers see Trump in a similar way that Mussolini's followers saw *Il Duce* ("the leader')—as a people's warrior who has raised the MAGA flag to symbolize the "Real America." Not by chance, Trump uses the handle "@realDonaldTrump" on his Twitter account.

TRUMP AND MUSSOLINI

Mussolini was a master wordsmith, who manipulated the mindset of Italian society with his own bricolage discourse—repeating the same slogans, clichés, and catchwords ritualistically. He railed against the politically correct speech of the usual suspects of his era—academics, liberal politicians, and intellectuals. His style of discourse resonated with his base, who felt that he talked directly to them, not about them.

As Ruth Ben-Ghiat eloquently argues in an *Atlantic* magazine article, this type of discourse has always been the style adopted by despots.[44] Mussolini fooled everyone when he came onto the political scene with his earthy language, setting himself apart from the intelligentsia of the era, tapping into common people's belief that liberal intellectuals looked down on anyone who did not talk or think like them. He founded Fascism as an "anti-party" just after World War I. He was seen as an outsider who came forward to drain Italy's political and social swamp. He challenged the extant politics of the nation, aiming to restore Italy to its purported greater past. His rabble-rouser followers took up his cause enthusiastically, terrorizing Italy with their own incendiary rhetoric and violent protests. Mussolini was a charismatic leader who trusted no one outside his own family, whom he put into positions of power after taking over the government. His rise to prominence was bolstered by blaming the insipid views of the liberal elite for the chaos and the crime rate, which he claimed were constantly rising in Italy.

Trump is a Mussolini clone. He also appointed his family to the government because of his distrust of others. He promised that, if he was elected, "The crime and violence that today afflicts our nation will soon come to an end,"[45] echoing Mussolini. As author B. Joey Basamanowicz emphasizes in his book, *Believe Me: 21 Lies Told by Donald Trump and What They Reveal about His Vision for America*,[46] America may be resurrecting a Mussolini-type era, whereby Trump's empty promises, revisionist histories, and baseless attacks on political and media opponents are defiling everyday discourse and ripping democratic institutions apart, in the same way that Mussolini's discourse did to Italian society.

Like Mussolini, Trump has created the false sense of reassurance in his followers that he would be able, *alone*, to solve problems connected to jobs, immigration, and health care. When Trump spoke to states bordering Mexico during the presidential campaign and whose citizens were affected by immigration, he proclaimed at various rallies that he would build a wall to keep illegal immigrants out of the United States. When he spoke to states whose citizens had grown weary of the liberal establishment he would promise to "drain the swamp." He knew, in effect, what type of discourse worked with a specific audience, using polling research to shape his talking points. In each rally speech, Trump pretends to be what that particular audience wants him to be. He is a modern-day Tartuffe—the fictional character in Molière's 1664 play, *Tartuffe*, subtitled, *The Imposter*. The parallels between Tartuffe and Trump are startling. In English, the word *tartuffe* is used to designate a hypocrite who exaggeratedly feigns virtue—a fact that is evident in Trump's support of moral causes, despite his profligate lifestyle.

Also like Mussolini, Trump portrays himself as the "savior leader" who is ready to free the people from the tyranny of liberalism and political correctness. James Pennebaker, an American social psychologist, found that pronoun use reveals what is in the mind of a speaker.[47] An analysis that I made of fifty tweets chosen at random and recordings of five rallies indicates that Trump never uses the pronoun *we* to refer to the government that he heads; rather, he uses *I*, setting up the image of a government that revolves around him, proclaiming himself to be the only true leader of the people, and calling those who oppose him "losers," "haters," and "lapdogs." Notice his tactical use of the *I* pronoun and reference to his unique intelligence in the following tweets:

> Sorry losers and haters, but my I.Q. is one of the highest—and you all know it! Please don't feel so stupid or insecure, it's not your fault.[48]

> I am the BEST builder, just look at what I've built.[49]
>
> Many people have said I'm the world's greatest writer of 140 character sentences.[50]
>
> I will be the greatest job-producing president in American history.[51]
>
> I will be the best by far in fighting terror.[52]
>
> Actually, throughout my life, my two greatest assets have been mental stability and being, like, really smart.[53]

This type of self-adulation is similar to Mussolini's. Even the name Mussolini adopted for himself, Il Duce, smacked of self-aggrandizement. It comes as no surprise to find that Trump has admired Mussolini, even though when confronted with this fact, he twists his response in dissimulative fashion. During an NBC *Meet the Press* interview, Trump defended a Mussolini quote he had retweeted. When challenged to explain why he wanted to be associated with a Fascist dictator, Trump cunningly replied as follows:

> No, I want to be associated with interesting quotes. And people, you know, I have almost 14 million people between Instagram and Facebook and Twitter and all of that. And we do interesting things. And I sent it out. And certainly, hey, it got your attention, didn't it?[54]

Trump sees himself as an American Duce. He has developed a whole lexicon of personal insults, as did Mussolini, which will be discussed in more detail later. These are intended to belittle opponents, so that he, the leader, can rise above everyone. During the primaries leading up to the 2016 presidential campaign, he called candidate Jeb Bush "low energy Jeb,"[55] Marco Rubio "little Marco,"[56] and then Hillary Clinton, his adversary in the campaign, "Crooked Hillary,"[57] all alluding to ascribed weaknesses in character or appearance. He called undocumented Mexicans "rapists and criminals"[58] and women "dogs."[59] These are not just insults; they are part of Trump's promotion of a persona who shows bravado by defying politically correct speech. Like Mussolini, he aims to destroy the liberal establishment by pitting himself against their nonsensical rules of hypocritical decorum that so many have come to loathe. Even his refrain of "draining the swamp" is modeled on a phrase used by Mussolini—*drenare la palude* ("drain the swamp"), which Mussolini used to rationalize his firing of over 35,000 civil servants when he came to power, as noted by Madeleine Albright (Obama's secretary of state) in her 2018 book, *Fascism: A Warning*.[60] Mussolini knew that with this metaphor he could justify the elimination of his enemies, convincing his followers that his opponents were part of an effete bureaucracy that was purportedly strangling Italy. As journalist John Kelly observes, the same phrase, "drain the swamp," appeared in a Wisconsin newspaper in 1903: "Socialists are not satisfied with killing a few of the mosquitoes which come from the capitalist swamp; they want to drain the swamp," which were the words uttered by a Social Democratic Party organizer.[61] The same expression has actually been used by politicians on the left and the right, including Pat Buchanan and Nancy Pelosi.

It is not possible to psychoanalyze Trump's understanding of the term, nor from which source he adopted it. Nevertheless, the parallel with Mussolini's metaphor is remarkable. Mussolini actually drained real physical swamps, using the phrase *drenare la palude* to acknowledge this real accomplishment. But he used it as well as a metaphor for the larger political and social swamp that he aimed to drain.

ALIENATION

The question of why a masterful liar-prince can conquer the minds and hearts of people through mendacity is *the* central one in any discussion of the Art of the Lie. One likely reason is that people occasionally feel marginalized and even ostracized by the society in which they live. There may be other reasons (of course), but this is a central one, as Machiavelli certainly understood long before the contemporary psychological research on alienation and marginalization. Already in the Renaissance, he understood that this feeling presents an opportunity for the liar-prince to promise a restoration of a sense of belonging—by any means possible, ethical or unethical. This has actually happened throughout history. A classic example is the French Revolution and the overthrow of the Bourbon monarchy in France in the period between 1789 and 1799. The crisis started with a meeting of the legislative assembly in May of 1789, when French society was undergoing a serious economic crisis. In July of that same year, the people stormed Bastille prison. The revolution became increasingly violent and ruthless under the leadership of the Jacobins and lawyer Maximilien de Robespierre. The execution of Louis XVI in January of 1793 was followed by Robespierre's so-called Reign of Terror, which failed to produce a stable form of republican government and was eventually overthrown by Napoleon in 1799. The people who rose up against the monarchy had become disenchanted with the aristocracy, feeling dispossessed and disparaged by its members. The latter sentiment came to be symbolized by a purported (but unverified) statement by Marie Antoinette, "Let them eat cake," which she supposedly uttered upon learning that the peasants did not even have bread to eat.

Psychologically, alienation is defined as a feeling of isolation from society, experienced by people who believe that society is unresponsive to their emotional needs.[62] The term was originally coined by Karl Marx to describe the sense of estrangement that he assumed working-class people (the proletariat) felt in a capitalist system.[63] But French social theorist Émile Durkheim suggested that alienation stemmed not from a particular type of polity, but rather from a loss of moral and religious traditions in a secularized and materialistic world.[64] He coined the term *anomie* to refer to the sense of irrational purposelessness that may arise in such a society.

Anomie may have been a subtle factor in the rise of Trump, who understood strategically that religious groups, such as white American evangelicals, felt an urgent need to restore their own model of morality to America. The religious right instantly saw Trump as the one who would make the restoration realizable politically, ignoring his morally checkered life.[65] Cleverly, Trump took their side on moral issues, assuring legislation to revisit abortion, promising to appoint federal judges and Supreme Court justices to overturn previous liberal judgments. In other words, many white evangelicals saw Trump as a vessel who would adopt their religious agenda and implement it, no matter what opposition he faced. (This topic will be discussed in more detail in chapters 3 and 7.) Before Trump's rise to power the mass media hardly paid attention to the religious right. Evangelicals saw the liberal media as promoting a secular worldview and moral relativism, while at the same time looking down implicitly and superciliously on their own religious views. They saw Trump as the person who would champion their religious ideology, no matter his philandering past lifestyle.

EPILOGUE

"Divide and conquer" is an expression that Machiavelli used in another one of his books, *The Art of War*, so as to suggest that the most effective strategy for conquering people is to divide the forces of the enemy, break up existing power structures, and stir up rivalries within the populace.[66] This can be done in military ways, of course, but the best approach is through the Art of the Lie—a tactic that Mussolini and Trump certainly understood before coming to power. Trump's divisive language is effective because it crawls surreptitiously into the subconscious, below the filters of conscious reflection, to produce images of those who are outside MAGA culture as "invaders" and "aliens," as Trump has called nonwhite immigrants. During the presidential campaign, he started to avoid heterogeneous crowds because of the chances of having to face unwanted conflict, talking exclusively to his base, which, he knew, would be enthusiastically supportive of his rallying cries against America's enemies—from within and without. In such a friendly atmosphere he could openly attack liberals and immigrants with brutal language. As George Orwell so insightfully put it, "Everyone believes in the atrocities of the enemy and disbelieves in those of his own side."[67]

Trump's followers are firmly in his camp, willing to accept anything he says, perceiving his discourse style as a primary tactic for fighting enemies in the ongoing "cold civil war," as the renowned journalist Carl Bernstein has characterized the dangerous state of affairs created by Trump: "We are in a cold civil war in this country. These two events, both the Mueller investigation and the Kavanaugh nomination, are almost the Gettysburg and Antietam, the absolute central battles of this cold civil war."[68] As journalist Ben Yagoda points out, the same phrase was used before Bernstein, in a 1950 *New York Times* article by one of its editors, Delbert Clark, in reference to the climate of fear and intimidation that McCarthyism had brought about in that era.[69] Discontented with America's liberal approaches to race, ethnicity, and morality, Trump's followers see the cultural war in terms of moral privilege. For this reason, his allies defend his stances and policies, no matter what they are, because they see themselves as his ersatz soldiers in this war. Trump's speeches are laden with dog whistles, symbolic metaphors, and bellicose slogans that fuel the will to fight the deep state.

Trumpian discourse has been promoted and reinforced by the alt-right social media universe. Even Trump's lies about his own lies are seen as strategic. This is why, as political commentator David Frum so aptly puts it, "Trump lies without qualm or remorse. If necessary, he then lies about the lie."[70] This climate of linguistic warfare engenders a need to attack the attackers, preempting the ability of the latter to be effective. "Head them off at the pass," was a cliché used in Hollywood cowboy movies of the 1940s and 1950s; it encapsulates Trump's counterattack strategy perfectly. His counterlies are so effective that they neutralize any potential retort or effective repartee. As a result of the silence that he has brought about, his lies gradually become accepted truths by his followers and apologists. As French writer Marcel Proust so fittingly observed, "Time passes, and little by little everything that we have spoken in falsehood becomes true."[71]

In his 1922 book, *Public Opinion*, American journalist Walter Lippmann argued that the growth of mass-media culture had a powerful direct effect on people's minds and behaviors.[72] His claim rings especially true today. Without the alt-right social media, Trump would likely not have become president. The cold civil war is fought more in cyberspace than it is in real space, with conservative media personalities constituting Trump's army generals. The American scholar Harold Lasswell foresaw the media's effect on belief systems nearly a century ago, suggesting in his 1927 work, *Propaganda Technique in the World War*, that the mass media affected people's politics, family relations, and general outlooks.[73] As Ralph Waldo Emerson also observed well before the age of electronic media, "Every violation of truth is not only a sort of suicide in the liar, but is a stab at the health of human society."[74]

The internet is especially critical in spreading the mind fog in which Trump thrives. Social media now wield great influence in politics. Issues of great concern are no longer restricted to discussions in the editorial pages of print newspapers, but through Twitter, Facebook, and other social media. As media scholar Sherry Turkle notes, in a media-saturated environment we are forgetting that, in the end, the face-to-face conversation is still the most effective discourse medium of all:

> But what do we forget when we talk through machines? We are tempted to forget the importance of face-to-face conversation, organization, and discipline in political action. We are tempted to forget that political change is often two steps forward and one step back. And that it usually takes a lot of time.[75]

To this assessment, I would add that political activism must eventually come from action in the real world, and especially from critical thinking engendered by reasoned arguments, not by slogans and counterslogans. Without this, there is no antidote to the machinations of the liar-prince, only exasperation and frustration. Aristotle defined rhetoric as "the faculty of discovering in every case the available means of persuasion."[76] He emphasized two main methods to counteract the negative effects of persuasion: truthful discourse and the use of logic to argue matters of importance. These are still the best remedies to neutralize the tactics in the Art of the Lie.

The remainder of this book aims to deconstruct this Art by discussing its various manifestations, present and past. In the preinternet era, writer Norman Mailer issued the following warning: "Each day a few more lies eat into the seed with which we are born, little institutional lies from the print of newspapers, the shock waves of television, and the sentimental cheats of the movie screen."[77] His warning is particularly relevant in the current age of the internet.

2
ALTERNATIVE FACTS

Everyone is entitled to his own opinion, but not to his own facts.

—attributed to Daniel Patrick Moynihan (1927–2003)

PROLOGUE

One of the most remarkable controversial public statements made in 2017 (which quickly spread throughout the mainstream media and became a meme in social media) was the one uttered by Kellyanne Conway, a Trump consultant and ardent ally, during a *Meet the Press* interview on January 27 of that year. When asked to explain Trump's false claim about attendance numbers at his inauguration, which he contended were much larger than they actually were—a claim repeated by Sean Spicer, Trump's press secretary at the time—she called it an "alternative fact." Her clever evasive statement was critiqued acerbically by those in the liberal media but espoused and praised by the radical conservative media as well as by Trump's fans and supporters, who saw Conway's phrase as encapsulating the *esprit* of the unorthodox language that Trump utilized during the campaign and in his presidency—a language seen as an "alternative" to the speech of those who inhabit the deep state—the supposed group of elite intellectuals, liberal politicians, and academics who lurk behind the scenes in American society, surreptitiously shaping the laws, policies, and politically correct discourse practices of the nation.

Conway's phrase was, upon closer scrutiny, a textbook example of Newspeak—the term coined by George Orwell in his novel *1984* (published in 1949) to designate the type of roundabout language used in a bleak, fictional totalitarian society named Oceania that was designed to create and maintain doubt and uncertainty.[1] The key feature of Newspeak is ambiguity, whereby the meanings of words and phrases can never be pinned down to any normal set of meanings. In their literal or concrete senses, words refer to some aspect of experience or to concepts based in, or derived from, the real world. In Newspeak, words do the opposite; they sever the normal connection of meanings to the real world and evoke instead an "alternative reality." As a result, no one knows what words really mean, or, more cynically, if objective reality as commonly understood even exists. This kind of verbiage allows Machiavellian manipulators to get away with saying virtually anything, including making claims that are empirically demonstrable as false or fake. Orwell defines Newspeak as a restructuring of orthodox vocabulary and grammar through the strategy of ambiguity, intended to allow those in power to keep the populace constantly in a mind fog that obstructs the clarity of thought necessary to initiate an uprising against the state. Semantic ambiguity is a trick in the Machiavellian liar's bag of verbal illusions; phrases such as "alternative facts" generate uncertainty and vagueness, disconnecting words from their normal meanings. As a result, the possibility of critical thinking is diminished. In Orwell's novel, the one who controls the meanings of words and what kinds of messages can be created with them is called Big Brother. He watches and monitors every utterance citizens make in order to detect any signs of unrest or nonconformity through any violation of the rules of Newspeak.

Big Brother's aim is to snatch truth from language and give it over to the Ministry of Truth, where alternative facts are manufactured. There, old news articles are rewritten with Newspeak so as to make them reflect the Ministry's version of reality and the old articles are thrown down a "memory hole" to be burned. The Ministry of Truth ensures that the party's version of truth is the only one. Orwell describes this state of affairs as follows:

> You believe that reality is something objective, external, existing in its own right. But I tell you, Winston, that reality is not external. Reality exists in the human mind, and nowhere else. Not in the individual mind, which can make mistakes, and in any case soon perishes; only in the mind of the Party, which is collective and immortal. Whatever the Party holds to be truth is truth. It is impossible to see reality except by looking through the eyes of the Party.[2]

Conway's phrase is taken directly from a page in the dictionary of Newspeak. The MAGA narrative is also taken from this dictionary. No one can pin down what Trump means by it. It is not, as Orwell put it above, something objective, external, existing in its own right. It is allusive and suggestive. So too are virtually all of Trump's slogans, such as "fake news" and "enemy of the people," which, when used over and over, draw many into their snare, encouraging them to live in the alternative reality that they circumscribe. In a word, such Orwellian language is a powerful ploy in the Art of the Lie. It allows the master liar-prince to create his own version of reality, through which he can easily manipulate people's minds by denying truth to words.

This chapter will focus on this ploy. As Orwell certainly knew, once people accept the language of alternative facts as part of discourse, the way they process normal speech and meaning is obstructed. This is a remarkable phenomenon that defies psychological explanation, and may require a reconsideration of the ancient concept of *mythos*—a type of language that is based on allusion rather than reference. Alternative-fact language plays ingeniously on our deeply rooted sense of *mythos*, as will be discussed below.

A FALSE EQUIVALENCY

It should be mentioned at the outset that some of the defenders of Conway's phraseology have claimed that the kind of abstruse, incoherent language she used was, ironically, invented by liberal academic elites themselves, stemming from the morally bankrupt movement known as "postmodernism." This is a false equivalency, though. Postmodernism was hardly an attempt to restructure language in an Orwellian sense, nor was it a mode of discourse within a political movement; it was an intellectual movement that has dissipated into the academic ether, having left no lasting traces in ordinary speech. So, before tackling the issue of Newspeak, a brief rebuttal of this critique is called for here.

Conservative writer and political thinker Ralph Benko is one of those critics who sees a connection between Trumpian alternative-fact language and postmodernism, as can be seen in the following statement:[3]

> The left is engaged in an all-out war on Trump and his supporters. One of its weapons is to attack declarations as "Fake News." Prominent journalists had a field day with Trump Counselor Kellyanne Conway for coining the phrase *alternative facts* in defending Sean Spicer's observations about the crowd size at Trump's inauguration. The left pioneered what it now criticizes. . . . As it happens, though, the left laid the foundation for "alternative facts." That's an artifact of a worldview that it pioneered. It condemns this as pernicious only when adopted by populist conservatives. What's really going on? Postmodernism, that's what.

To reiterate, postmodernism was not a brainchild of leftist politicians, nor was it influential in shaping mainstream politics in any enduring way. At best, it encouraged a type of discourse of blandness designed to be inoffensive that was bandied about in academia for a while; ironically, it was constantly under attack by the academics themselves as being too abstruse and self-serving. What defenders of Conway are confusing is postmodernism with political correctness—a theme that will be discussed in detail in chapter 6. And they are likely doing this deliberately, rather than accidentally.

Postmodernism took a foothold in several academic disciplines in the 1980s and 1990s—mainly in literary studies, semiotics, and popular culture studies. The term was coined originally by architects in the early 1970s to designate an architectural style that aimed to break away from the preexisting modernist style, characterized by indistinct boxlike skyscrapers and apartment buildings that had degenerated into sterile and monotonous structural formulas. Postmodern architects called for greater individuality, complexity, and eccentricity in design, along with the use of architectural symbols with historical value. Shortly after its introduction into architecture, the term caught on more broadly, adopted by many in the arts and humanities. This is somewhat reductive, but it nonetheless captures in a nutshell what the postmodern movement aimed to achieve—unsuccessfully in my opinion.

The probable reason why the term became fashionable in some quarters was because it articulated a fomenting reaction against a rigid form of critical analysis that had been the standard in universities. Many saw this new form of "open" criticism as a welcome relief from the stiffness, but very few actually allowed it to become dominant outside of humanistic disciplines. It was found rarely (if at all) in disciplines such as psychology, sociology, and linguistics. And even those who were favorable to it at the time acknowledged that, overall, it tended to veer off into self-serving verbiage. Some have connected postmodernism to the New Left political movement. But this is also inaccurate. The New Left movement was "New" only in relation to the "Old Left" that was guided by Marxist ideas. New Left academics and politicians demanded sweeping and fundamental changes to major institutions, public and private, aiming to put an end to such injustices as sexism, poverty, racial discrimination, and class distinctions. Members of the New Left favored civil disobedience, which occasionally led to clashes with the authorities. But these were never widespread, and they never truly threatened the political status quo in any effective way. More to the point of the present discussion, the radical New Left movement should not be linked to postmodernism, as some conservative critics attempt to do.

The objective of the postmodernists was not to control or restructure people's thoughts, as it so obviously is by those who use expressions such as "alternative facts"; it was just the opposite—to allow thoughts to be free from the yoke of tradition and the overreliance on formulas. It ended up having no lasting effect on the mainstream humanities and social sciences, but it did raise questions about the nature of knowledge and language itself. Like previous intellectual movements, it has run its course. To label Conway's left-wing critics as postmodernists is to beat a dead horse.

DOUBLETHINK AND DOUBLESPEAK

We all harbor deeply ingrained beliefs that color and filter the information we glean from conversations and other kinds of social interactions. If we are honest, we would admit that we never truly interpret the "facts" that are in the information objectively, free of our unconscious biases and beliefs. However, in most normal speech situations, we do attempt to make sense of the truth of the matter at hand. Of course, the skilled liar can easily manipulate our attempts by restructuring the meaning of the words that carry the information. This is done in many ways. The one that is of interest in this chapter is the strategy of denying words their normal meanings regularly and systematically. Conway's phrase of "alternative facts" has no real meaning; it is equivocal at best and deceptive at worst, allowing her to skirt around the false claim made by Trump about audience size through a clever play on words. It creates doubt in people's minds leading to the sense that it may have plausibility.

This ploy is a part of an overall strategy of restructuring language in such a way that allows for the control of meaning through the creation of doubt or uncertainty. Trump's followers, allies, and apologists use this linguistic strategy constantly in their "talking points" during media interviews, creating confusion with cleverly restructured words and phrases, projecting discussions and debates into a miasma of doubts and uncertainties that inhibit clarity of thought. The goal is to throw the truthful words of opponents down the "memory hole" to be incinerated. The language devised strategically by Conway and other Trump followers is, in other words, a case of Orwellian doublespeak, a language that intends to evade uncomfortable facts by inducing a form of doublethink, a state of mind that accepts mutually exclusive ideas as both possible. In Orwell's novel, anyone who identifies any contradiction in doublespeak is immediately captured and subjected to disciplinary action.

As media analyst Edward S. Herman has cogently argued in his book, *Beyond Hypocrisy*, doublespeak is nothing more than a skillful utilization of lying:[4]

> What is really important in the world of doublespeak is the ability to lie, whether knowingly or unconsciously, and to get away with it; and the ability to use lies and choose and shape facts selectively, blocking out those that don't fit an agenda or program.

Doublethink is a state that accepts contradictory beliefs or ideas as plausible, being unaware of any contradiction. Orwell explains that while it may appear absurd at first, over time it develops its own form of cogency that people feel is just as valid as any conceptual system. He goes on to suggest that this occurs because of peer pressure among users—a situation that is evidenced by the fact that Trump's followers seem to encourage each other constantly to hold the line and remain faithful to the cause. Doublespeak thus forms a kind of in-group code with its own set of "alternative truths." Incidentally, the word *doublespeak* does not appear in Orwell's novel; it was coined after its publication to designate the kind of language he describes in the novel whose aim is to deliberately disguise or distort word meanings and to manufacture consent—a term coined by Herman and Chomsky in reference to the manipulation of language to gain a consensus on political ideology.[5]

Doublespeak can be characterized as a language put together through combinations of "verbal chemicals" that do not form any normal molecular blend (to extend the metaphor). Rather, they are put together as "antagonistic" elements in order to distort meaning. The words *alternative* and *facts* resist amalgamation in normal semantics; but in doublespeak their combination is designed to produce doublethink. The wordsmiths of doublespeak are linguistic alchemists who know how to combine incompatible semantic elements to generate their alternative realities.

Orwell saw this kind of language as the quintessential strategy of mind control used in totalitarian states. It weakens minds through contradiction and the denial of objective truth that, together, allow the dictator (Big Brother) to exact conformity to his own way of seeing things among the populace. The following excerpt from a speech given by Stalin to the Sixteenth Congress of the Russian Communist Party in 1930 encapsulates the political strategy underlying doublethink perfectly:[6]

> We are for the withering away of the state, and at the same time we stand for the strengthening of the dictatorship, which represents the most powerful and mighty of all forms of the state which have existed up to the present day. The highest development of the power of the state, with the object of preparing the conditions of the withering away of the state: that is the Marxist formula. Is it "contradictory"? Yes, it is "contradictory." But this contradiction is a living thing and wholly reflects the Marxist dialectic.

In his insightful book, *Beyond Hypocrisy* (cited above), Edward Herman argues that our minds are extremely vulnerable to such verbal alchemy, more so than they are to outright intimidation and confrontation, because vagueness of meaning obscures and may even obliterate our basic assumptions about reality, leading to a sense that any combination of words and distortions of meanings produce sense.[7] If such manipulation spreads broadly among a certain group, it can lead to the communal acceptance of contradiction as a "living thing," as Stalin cleverly put it.

In Orwell's novel, Newspeak was the primary means for establishing a social order that could be controlled by Big Brother, who made sure that the articulation of contrary ideas was blocked through a control of word meanings.

In addition to an alchemical blending of words, so to speak, doublespeak utilizes negation as a central strategy. For example, the word *bad* is revised as "ungood" or "doubleplus ungood," if something is particularly bad. This is a sinister strategy that blocks the use of effective negative arguments (repudiation, refutation, or rebuttal) against the state. It generates doublethink, which is described by Orwell as follows:[8]

> To know and not to know, to be conscious of complete truthfulness while telling carefully constructed lies, to hold simultaneously two opinions which cancelled out, knowing them to be contradictory and believing in both of them, to use logic against logic, to repudiate morality while laying claim to it, to believe that democracy was impossible and that the Party was the guardian of democracy, to forget whatever it was necessary to forget, then to draw it back into memory again at the moment when it was needed, and then promptly to forget it again, and above all, to apply the same process to the process itself—that was the ultimate subtlety: consciously to induce unconsciousness, and then, once again, to become unconscious of the act of hypnosis you had just performed. Even to understand the word—*doublethink*—involved the use of doublethink.

The overriding objective of doublespeak is to generate an "alternative reality" that consists of "alternative facts," which "induce unconsciousness" and forgetfulness of actual facts, as Orwell put it. As Stalin and other mind managers knew, this is one of the most effective of all the mind-controlling strategies in the Art of the Lie. During a speech he gave in Kansas City in July 2018, Trump made the following statement: "What you're seeing and what you're reading is not what's happening."[9] This is taken right out of Big Brother's playbook: "The party told you to reject the evidence of your eyes and ears. It was their final, most essential command."[10] It intends to assign truth to only one source—Trump. No other sources can be trusted since they are "enemies of the people," an expression that originated, not surprisingly, with Stalin.

Examples of doublespeak abound in Trump's tweets, rally speeches, and statements, such as the false crowd size claim above, which he concocted to portray himself as not only a populist leader but a popular one as well. Strategic attacks on truth-tellers is also part of the overall doublespeak ploy, as can be seen in Trump's frequent assertions denying climate change, which are indirect attacks on the "liberal elite" whom he claims are lying to the people for their own purported self-serving objectives. Below is a typical example. Significantly, the fact that hundreds of thousands "liked" the tweet is a sure sign that the doublespeak strategy is an effective one indeed, showing that this kind of language seeps surreptitiously into the unconscious part of the mind, altering its views of reality:

> In the East, it could be the COLDEST New Year's Eve on record. Perhaps we could use a little bit of that good old Global Warming that our Country, but not other countries, was going to pay TRILLIONS OF DOLLARS to protect against. Bundle up![11]

Doublespeak is based on the restructuring of the vocabulary of a language so that any word in it can be transformed to serve the objectives of the state. Orwell divides it into three categories—A, B, and C. The "A Vocabulary" contains common words and phrases that, as Orwell says, are "for such things as eating, drinking, working," and so on. These are few in number and must be constrained to their literal meaning. They must be blocked from accruing nuances of meaning, which would increase the semantic versatility of the items in Vocabulary A considerably and thus turn them potentially into dangerous verbal weapons that can be used against the state. A-words must be regulated constantly, tying them down to their literal or primary meanings constantly, so as to avoid the danger that they can be used to invoke ideas or feelings that would be detrimental toward the state.

The "B Vocabulary" contains words with controlled political and ideological meanings, tailored to engender blind acceptance of the government's doctrines through techniques that generate ambiguity, obfuscation, vagueness, ambivalence, and double entendre. For example, "goodthink," which means roughly the same thing as "orthodoxy," produces an ambivalent sensation enticing the speaker to imagine a world that is correct and righteous, while the latter term ("orthodoxy") may initiate doubts about the restrictive effects of conformity and suggest independence of mind, by implication or connotation. This part of the vocabulary consists mainly of compound or compressed words, which are intended to achieve conceptual obfuscation: the phrase *Thought Police* is compressed into "thinkpol"; *Ministry of Love* becomes "miniluv"; and so on.

The "C Vocabulary" is made up of words that relate specifically to science and technical disciplines, ensuring that such knowledge remains segmented and specialized. This means in practice that scientific knowledge is restricted to a few and thus to be kept away from the masses who might use it against the state. There can also be no single word for *science*, since this would by itself entail thinking about the world in terms of new knowledge and discovery.

Overall, the doublespeak lexicon is designed to block critical thinking, lucidity of thought, and overall reasoning. It is also a powerful form of subtle hypnosis that it brings about through catchwords used over and over, producing a spellbinding effect. Trumpian catchphrases such as *Great*, *Sad*, *Wrong*, *Believe me*, and more are examples of how he uses this feature of doublespeak effectively to produce his own form of hypnosis on his followers. Not surprisingly, Mussolini also used stock phrases, such as *State ownership*, *Anarchist*, and *Relativism*, in his emotionally charged speeches:[12]

> State Ownership! It leads only to absurd and monstrous conclusions; state ownership means state monopoly, concentrated in the hands of one party and its adherents, and that state brings only ruin and bankruptcy to all.

Every anarchist is a baffled dictator.

> If relativism signifies contempt for fixed categories and those who claim to be the bearers of objective immortal truth, then there is nothing more relativistic than Fascist attitudes and activity. From the fact that all ideologies are of equal value, we Fascists conclude that we have the right to create our own ideology and to enforce it with all the energy of which we are capable.

Those who frequented Mussolini's rallies were mesmerized by his slogans, reacting in unison to his bluster by shouting consent and approval frenetically. The meaning of this kind of Orwellian atmosphere was captured by Apple's brilliant 1984 TV commercial, which was shown on January 22, 1984, during the third quarter of Super Bowl XVIII, and directed by Ridley Scott. The commercial achieved two things at once—it introduced the new Mac computer at the same time that it treated the dangers of a world of Orwellian conformity based on a language of catchphrases and slogans.

Scott cleverly used the medium of a commercial to issue a cautionary tale about doublespeak and doublethink. The number *1984* appears at the start of the commercial, as a horde of shaved-head, expressionless men, in prison-style uniforms and boots, march mindlessly toward a gigantic TV screen where a Big Brother shouts meaningless Newspeak platitudes at them. The men stare at the screen in a zombie-like state. Then, out of nowhere, a blonde, attractive, athletic woman appears in a white jersey and red shorts running toward the men, pursued by a group of storm troopers (an obvious Gestapo allusion). She enters the room, hurling a sledgehammer at the television screen that, as a result, explodes.

The men remain seated, open-mouthed and dazed, ready to come out of their stupor.

Although the objective of the commercial was to introduce the new Mac computer, its symbolism and psychological implications were unmistakable—the only way to break the conformity-inducing effects of doublespeak is through the leadership of a "goddess," who brings forth freedom from mental slavery. Arthur Asa Berger puts it perceptively, as follows:[13]

> The blonde heroine, then, is an Eve figure who brings knowledge of good and evil, and by implication, knowledge of reality, to the inmates. We do not see their transformation after the destruction of the Big Brother figure—indeed, their immediate reaction is awe and stupefaction—but ultimately we cannot help but assume that something will happen and they will be liberated.

The apparition of the woman "humanizes" the mindless throng, as they open their eyes, coming out of their mental cocoons. Perhaps there is no better counterattack to Big Brother and his ability to control minds than to allow a goddess to bring forth a new sense of life and freedom—a theme that goes back to antiquity and myths such as the Gaia one, as will be discussed in the concluding chapter.

ALTERNATIVE HISTORY

Every society develops narratives of its historical origins; these allow the members of a collectivity to interpret the raison d'être of their institutions, beliefs, laws, symbols, customs, and so forth. Historians are de facto "truth-makers" because they are the ones who take the events of history and assemble them into a narrative that provides a sense of meaningful connectivity to the past. So, the way a historical narrative is written or told shapes the way people envision their roots and how they come to view the world. It is little wonder, therefore, that in Orwell's novel, historical narration is controlled by the Ministry of Truth, which shapes the facts to fit into the party's worldview. The historical records are preserved and written in doublespeak. This type of history can be termed *alternative history*. It can be defined as a historical narrative devised to impart and establish alternative facts through doublespeak. All conspiracy theories are alternative histories.

The protagonist of Orwell's novel, Winston Smith, works in the Records Department of the Ministry of Truth. His job is to revise historical records in order to ensure that the past conforms to the party line, deleting any inconvenient facts perpetrated by so-called unpersons—that is, by those who oppose the state and whose memory of the facts must be denied by "vaporization." The Ministry's main objective is to obfuscate real events, including those that are linked to personal histories. The description of Smith's birthday is a case in point: "It was a bright cold day in April, and the clocks were striking thirteen."[14] This implies that the factual date of his birth remains uncertain, leaving him in a mind fog. All records of people and of historical events are encoded in a similarly vague and allusive manner, allowing the Ministry to control people's understanding of past events. As Orwell understood, alternative histories allow totalitarian regimes to spin conspiracy theories that shield them from effective opposition. As Chaim Shinar has perceptively observed, this very strategy was used by Stalin to silence the opposition against him, keeping people in a state of fear and uncertainty; Vladimir Putin uses the same tactic in order to maintain power, claiming that there is an international conspiracy against Russia by those who oppose its mission and goals, which are actually his own mission and goals.[15]

Creating alternative narratives is a key strategy in Trump's bag of Orwellian tricks. The deep state slogan is a case in point; it alludes to a conspiracy being contrived by a group of people who had the reign of power before he did, and who, behind the scenes, are plotting to remove him from power because they fear he will expose them. This craftily skewed doublespeak metaphor works emotionally for his most ardent followers because it claims to tell the "real story" of the overtaking of America by "un-American" liberals. Conspiracy theories work because they utilize alternative facts effectively to spin reality around and around. A classic example is that of criminal societies such as the Mafia, which distort history with their own form of doublespeak so as to legitimize their own existence socially (as will be discussed).[16]

The ancient Greeks divided language and thought—including the language used in writing history—into two main categories: *lógos* and *mythos.* This is admittedly somewhat reductive, but nonetheless accurate in outline form. The former referred to historical narration that was based on an interpretation of actual events, assembled into a sequential concatenation that aims to document how the present is implanted on the past; the latter referred to narrations that were based instead on beliefs, assembled in terms of events that are perceived to be part of a larger metaphysical reality. Historical narratives based on *lógos* aim to provide a rational interpretation of past events; mythic narratives provide instead an explanation of events in terms of their metaphysical implications.

Mythos was coined by Aristotle to describe the narratives of tragedies, defined more or less as the recurrent theme or plot structure that characterizes them. This is part of a human-versus-gods drama that unfolds in the imagination. The great tragedies of Aeschylus, Euripides, and Sophocles dealt with mythic history as means to understand human destiny. Most early myths thus have an important historical function—they provide stories that allow people to grasp the meaning of recurring themes in human life—good versus evil, life versus death, and so on. These themes are embedded into the human unconscious, leading to the development of beliefs that shape human consciousness.[17] Aristotle claimed that *mythos* may have indeed served an important psychological function, but it could also be easily manipulated because it was based on belief rather than rational understanding.[18] The deep state conspiracy narrative that American society has been destroyed by liberals and their political correctness schemes is an example of how *mythos* works psychologically. It cannot be demonstrated as true in any objective way; it can only be believed in the same way that mythic stories are. The same kind of technique was used by Mussolini, who identified liberals and intellectuals as the scourge of Italian society in his era. Conspiracy narratives allow any masterful liar to render the arguments of opponents ineffectual, since they are portrayed in the mythic conspiracy as the villains in the narrative. By labeling the mainstream media in America as the "enemies of the people," Trump has revived the same kind of mythic narrative used by Stalin. As Orwell so aptly put it, "Myths which are believed in tend to become real."[19]

Conspiracy narratives are powerful tools of mind control since they are designed to stoke deeply engrained beliefs. Indeed, when evidence showing them to be false is presented to the believers, they interpret such evidence as actual evidence of their truth, because the refuters are seen as nonbelievers or skeptics. As Michael Barkun has written, conspiracy narratives rely on three basic principles of *mythos*—nothing happens by accident, nothing is as it seems, and everything is connected.[20] This kind of mindset is a closed one that cannot be altered because truth is a "matter of faith rather than proof."[21]

Conspiracy narratives work psychologically because they are based on a strategic use of metaphor. A perfect example is Trump's metaphor of the "witch hunt," as he has characterized the Russia investigation into alleged collusion between his presidential campaign and Russia aiming to influence the outcome of the 2016 election in his favor. This type of language is highly allusive and charged with historical meaning, pointing to a tragic period of dangerous mind control in American history—the Salem witch trials in colonial Massachusetts in 1692 to 1693. There is little doubt that Trump's opportunistic phrase is intended to evoke a sentiment of false persecution, undermining therefore the validity of any prosecution that ensues from the investigation, since it would be seen by believers as verifying the conspiracy against him by the deep state. So, no matter what the investigation reveals, Trump's followers will still see it as a witch hunt.

Such narratives gain incremental credibility if they are repeated over and over, as they are, for example, through social media platforms, which have become the primary channels of dissemination of these narratives. The witch hunt mythology would never have gained believability without the radical conservative social media, which have raised it to the level of political chronicle. No countervailing argument, based on *lógos*, can ever penetrate the ingrained verisimilitude of the story, in which the persecuted victim, Trump himself, is seen as a martyr at the hands of the forces of the deep state.

As an aside, *mythos* by itself is not a negative aspect of human cognition. It is used to this day in children's stories, fables, and legends that allow adults to impart the meanings of ethics and morals. It is the manipulation of *mythos* that is at play in alternative narratives, not *mythos* itself. The false story that America is being "invaded" by hordes of reckless immigrants is an example of how Trump manipulates *mythos*. Like the Russian conspiracy narrative above, it is designed to evoke fears that those who come into a nation from outside are invaders and must be dealt with severely. As J. P. Linstroth cogently reminds us, Trump is tapping into a long-standing myth of nativism in America—namely, that immigrants pose a threat to "nativist" American culture, whatever that may be:[22]

> Toward the end of the 19th-century and at the turn of the 20th-century, many in the US promoted "nativism"—an all-white America where good jobs belonged exclusively to whites. This was the historical period known as the "Second-Industrial Revolution," the "Gilded Age," and the "Progressive Era"—a time of enormous economic transformation for the country through industrialization and urbanization.

A false mythology speaks directly to a group of people who harbor inner resentments, such as those who saw the Obama presidency as an elitist one that excluded them from its purview. One pundit who supported Trump during the election campaign made the following revealing statement on national television, paraphrased here through recollection: "We will be excluded no more; we mention race and we are called racist; we mention immigration and we are called xenophobic. This will stop under Trump (paraphrase mine)." The real achievement of the Machiavellian liar is to get people to notice him as their only way out of their sensed fears, resentments, and dilemmas. As in the ancient mythic stories, Trump emerges as the heroic figure to set things right in the world, despite his flaws—after all, the mythic heroic figures have tragic flaws. The invasion myth allows Trump to gain power over reality itself. To his followers, what Trump says is true, if *he* says it is. Trump is the only figure who appears distinct in the mind fog of alternative history—everyone and everything else is a blur.

Alternative conspiracy histories today have found fertile ground for dissemination, as mentioned several times, in cyberspace. The Web is now the main conduit for mythic storytelling and the making of legends. Ideas no longer spread primarily through print or by word of mouth but by internet memes and viral videos. As Richard Dawkins, the originator of the term *meme* long before the internet, claimed, memes are just as transferable to others as are genes.[23] Mythological and conspiracy narrative memes are particularly effective on susceptible people.[24] Cyberspace and its meme structure might be changing—or mutating—human understanding, taking us right back to a *mythos* form of consciousness where anything that appears in memetic form is likely to be believed.

A main objective of conspiracy narratives is to "mobilize passions," as Robert Paxton points out in his book, *The Anatomy of Fascism*.[25] The main passion mobilized is a sense of overwhelming crisis, based on a dread that the original "pure society" is under alien attack. The invasion narrative taps into this sense of dread, allowing Trump to highlight his leadership instincts over those of soppy and weak opponents. Similar invasion narratives are found in virtually all fascist and totalitarian regimes, as Paxton illustrates. Hitler always proclaimed the right of the chosen people to rule the world through his Aryan myth since they, and they alone, were the ones chosen by destiny to dominate others without restraint. This myth will be discussed in the next chapter.

Alternative narratives are truly Orwellian, allowing ruthless masters of the Art of the Lie to blame "others" for their own problems and, at the same time, to legitimize their own tales. They reveal a form of Freudian projection that is self-serving but ultimately destructive. Eventually, the Winston Smiths of the world will have to come to grips with reality, as they did in Italy under fascism and in Germany under Hitler.

RESTRUCTURING THE LEXICON

A key lesson to be learned from Orwell's *1984* is that it is ludicrously easy to manipulate human minds—it can be done much too simply through a clever restructuring of the lexicon and by spinning mythic stories. This suggests that "words do indeed matter," as the commonly used cliché implies. We process the meaning of words unconsciously and, as Orwell knew, this process can be altered or constrained by changing the connection between words and their meanings.

Orwell warned that the rise of totalitarianism was more likely to emerge when language was distorted to serve the machinations of the dictator, more so than any other factor, including military action. Newspeak was his way of arguing how easily this can be done. The suppression of free thought can be engineered simply by restructuring the lexicon and controlling the semantics of the items within it. Trump achieves a similar kind of "meaning control" by assigning to specific words and phrases a sense and mythic import that plays constantly on peoples' fears and resentments. Slogans such as the "deep state" and "MAGA" ones empower him to control meaning in an Orwellian way because they allow him to characterize those who oppose him as villains within a conspiracy. The slogans are, in effect, Ministry of Truth ploys that are designed to stoke the belief that Democrats, liberals, the media, and anyone else who criticizes him have destroyed society, at the same time that he, Trump, will flush them from the deep state to make things right and return to an idyllic American past (MAGA). This type of strategically coded language is meant to tap into fears and resentments, which Trump manipulates through such metaphors and slogans, repeated over and over, inducing a chantlike hypnotic effect on people's minds. The result is the crystallization of an alternative view of history that speaks directly to beliefs, not to facts.

The mind control that such language allows a Machiavellian liar to achieve is thus a shield against any opposing argument or evidence that he is a liar. Logical counterarguments are ineffectual because they are perceived as the words of the "enemy," and thus easily dismissed as fallacious or self-serving. As the liar-prince knows, this kind of control impels followers to shelter and protect him, almost robotically, from adversities of all kinds. It is a mindset that is tribal and typical of village-type congregations, as French writer Jean de La Bruyère pointed out in 1608:[26]

> The town is divided into various groups, which form so many little states, each with its own laws and customs, its jargon and its jokes. While the association holds and the fashion lasts, they admit nothing well said or well done except by one of themselves, and they are incapable of appreciating anything from another source, to the point of despising those who are not initiated into their mysteries.

As Aldous Huxley wrote, in-group savvy is a powerful motivating force in human behavior: "To associate with other like-minded people in small, purposeful groups is for the great majority of men and women a source of profound psychological satisfaction. Exclusiveness will add to the pleasure of being several, but at one; and secrecy will intensify it almost to ecstasy."[27] The language of alternative facts unites like-minded people who are brainwashed to go out of their way to constantly protect the leader from all adversity. Any perceived injustice against the leader by an ex-member of the group will be despised, as de La Bruyère so aptly stated. In Mafia culture, such a traitor is labeled a "rat." When Trump called his previous lawyer Michael Cohen a "rat," he was engaging in a similar form of Mafia slang. Trump and Rudy Giuliani (one of Trump's lawyers) wrote tweets that were clearly intended to intimidate Cohen from testifying publicly to Congress in early 2019, threatening to expose Cohen's father-in-law and wife as having carried out illegal activities in the past. As Emile Durkheim has suggested, this kind of speech is designed to ensure the "mechanical solidarity" of the group.[28]

Lexical restructuring is achieved in many ways, some of which have already been discussed, and others will be analyzed in more detail subsequently. A few examples of how it produces Orwellian effects will suffice here:

"Cuckservative." This is a blend of *cuckold* and *conservative*, used commonly by alt-right media pundits as an insult to anyone who sells out the political base. As in Newspeak, the creation of compound words is a major strategy in this kind of discourse, constituting its B Vocabulary. A term such as this one fragments meaning into bits and pieces that are assembled according to conservative beliefs.

"Total loser." This is used often by Trump to attack anyone who disagrees with him or says something negative about him. It is part of an attack strategy, as will be discussed subsequently in more detail. Expressions of this kind control the discourse flow because they block any effective counterattack by the "loser," since it can refer to anything by innuendo and thus cannot be directly negated. It also puts the attacker on the defensive, with little opportunity to engage in repartee.

"Bad hombres." This is Trump's moniker for illegal Mexican immigrants. As he said during the final televised presidential debate: "We have some bad hombres here and we're going to get them out."[29] It is a clever metaphorical play on the Spanish language while at the same time evoking an image of villains, looters, and miscreants that he uses to attack Hispanic immigrants. The term *bad hombres* was also used in Hollywood Western movies from the 1930s to the 1960s, with the meaning of outlaws.

To his opponents, the language used by Trump may appear to be outrageous and transgressive of norms, but to followers and allies it is part of a bricolage of meanings that tap into a reservoir of resentments and animosities against liberal approaches to governance. Through tweets, online talk shows, television cable channels, and more, any Winston Smith can establish alternative facts as real, without ever having to provide any empirical evidence that these actually exist.

The Orwellian restructuring of the lexicon strengthens group beliefs at the same time that it weakens the "enemy's" will to fight. Shutting down the US government in December and January 2018 to 2019 in order to get a "wall" to stop illegal immigration from Mexico is an example of how Trump aims to achieve this goal. The Democrats called it "taking the country hostage," while the alt-right pundits called it a wise and opportune way to solve the problem of immigrant "invasions." Trump's allies would refer to the wall solution as moral and right, thus projecting it emotionally into the larger alternative conspiracy narrative for the overthrow of the immoral deep state. The wall was a powerful metaphor that also tapped into the MAGA narrative. Trump constantly made up statistics to indicate the need for a wall, claiming the immigration situation at the Mexican border as a "human crisis." He has repeated such falsehoods over and over—a tactic that strengthens the resolve of believers. As Hitler wrote in *Mein Kampf*: "The intelligence of the masses is small. Their forgetfulness is great. They must be told the same thing a thousand times."[30] By embedding the same alternative narrative into the fabric of news reports, it gains strength. It also expresses a situation for followers in a concrete way, as media scholar Francesco Mangiapane has aptly observed:[31]

> Almost entirely absent [*from such news sites*] are posts that call for the reader to make an effort to interpret the post or call on their critical abilities. These sites take nothing for granted. Like in *telenovelas* of the past all ambiguity is cancelled out and the tendency is to guide the story through predictable, unproblematic scenarios.

Restructuring the lexicon has always been a tactic of propaganda. It is relevant to note that the term *propaganda* derives from the Latin name of a group of Roman Catholic cardinals, the *Congregatio de Propaganda Fide* (Congregation for the Propagation of the Faith) established by Pope Gregory XV in 1622 to supervise missionaries. Gradually, the word came to mean any effort to spread beliefs of a particular kind. It acquired its political meaning after World War I when journalists exposed the dishonest but effective techniques that propagandists had used during the war. In the early 1900s, Vladimir Lenin, who led the Communist Revolution in Russia, argued that propaganda works because it uses half-truths and slogans to arouse the masses, whom he considered incapable of understanding complicated ideas. Not surprisingly, the techniques of propaganda were adopted in 1922 by Mussolini, allowing him to establish a fascist dictatorship in Italy by "telling it like it is." Fascist propaganda promised to restore Italy to the glory of ancient Rome. As Mussolini derisively and effectively put it: "A nation of spaghetti eaters cannot restore Roman civilization."[32] In referring to education, he stated the following: "Fascist education is moral, physical, social, and military: it aims to create a complete and harmoniously developed human, a Fascist one according to our views."[33] Joseph Stalin, who led the Soviet Union in the late 1920s, used propaganda to crush all opposition. And in 1933, Adolf Hitler set up his Nazi dictatorship in Germany, imposing his own form of propaganda speech throughout society, stoking racist and xenophobic paranoia among the populace.

Needless to say, democratic governments have also used propaganda, but it would be a false equivalency to compare such usage with that of a Hitler, a Stalin, or a Mussolini. In 1953, the American government established the US Information Agency (USIA) to create support of its foreign policy. The Voice of America, the radio division of the USIA, broadcast entertainment, news, and American-based propaganda throughout the world. The government used the Central Intelligence Agency to spread covert propaganda against governments unfriendly to the United States, including those of the Soviet Union and the communist countries of Eastern Europe. The CIA also provided funds to establish radio networks called Radio Free Europe and Radio Liberty, which broadcast to communist countries. Without going further into the politics of propaganda, suffice it to say that in all cases, the control of meaning by restructuring the lexicon is the key tactic of propagandistic discourse, and this can be achieved through specific linguistic ploys, some of which have been discussed briefly in this chapter. There is no mind control without vocabulary control.

THE POWER OF BELIEF

In the beginning stages of human cultures, myths functioned as genuine explanations of the world, including how they originated (as mentioned briefly above). The Zuñi people of North America, for instance, claim to have emerged from a mystical hole in the earth, thus establishing their kinship with the land; Rome was said to have been founded by Romulus, who as an infant had to be suckled by a wolf, thus alluding to an innate fierceness that the Roman people believed they possessed; and the list could go on and on. Myths create a belief system that becomes the basis for a culture's institutions, including religious and family ones. Even today, as discussed, we resort to mythical storytelling for imparting knowledge of values and morals, initially to children.

By studying *mythos*, we can arguably learn a lot about how people develop a particular worldview and thus better understand the values and beliefs that bind members of groups together. Especially critical in this analytical framework, in the context of the present discussion, is the so-called eschatological myth, which aims to describe the end of the world. An apocalypse, such as a universal fire or a final catastrophic battle, is one of the most emotionally powerful of any eschatological mythology. To counteract the apocalypse, many cultures tell derived myths of the coming of a savior, called the *culture hero*, who will prevent the disastrous end.

Many white evangelicals in the United States have supported Trump despite his philandering ways because they see him in apocalyptic terms—as a culture hero, sent from above to set things right in the world. An interview on CNN in 2018 provided some evidence to back up this theory.[34] A group of white evangelical women was asked why they supported Trump in the election and continue to support him zealously as president. The common answer was that he was a "godsend," pointing to a picture of him displayed prominently on a fireplace, much like a religious icon. As a culture hero, Trump is "sent from above" as the destroyer of the enemies of the people, who are believed to be atheist intellectuals and politicians who are ruining the moral fiber of society. As a master Machiavellian deceiver, there is little doubt that Trump promotes the views of the religious right opportunistically, not only with his policies but also, and especially, with his discourse that involves keywords such as *life* (of the unborn), *family*, and, above all else, *religious freedom*.

Such groups in Trump's base believe, literally, that he has come to restore moral order by destroying the immoral chaos in which we were purportedly catapulted by relativism and liberalism, bunched together under the rubric of postmodernism (as discussed). According to the *Theogony* of the Greek poet Hesiod (eighth century BCE), Chaos generated Earth, from which arose the starry, cloud-filled Heaven.[35] In a later myth, Chaos was portrayed as the formless matter from which the Cosmos was created. In both versions, it is obvious that the ancients felt that Order arose out of Chaos. At an emotional level, Trump's rise to power, for some religious people, is seen as fulfilling a "theogonic" destiny, a restoration of Order by destroying the liberal-created Chaos. The underlying theme in this mythic narrative is that only a destroyer can restore the moral order, even if he is himself a sinner. This type of belief is consistent with some of the ancient catastrophe myths, whereby the godsend is a flawed character who will nonetheless guarantee salvation through his own power of destruction.

It is thus little wonder that Trump is viewed as a larger-than-life culture hero by some groups—a fact that virtually guarantees their permanent support for his leadership, no matter how he behaves in real life, or what kind of profane and vulgar language he uses. As a sinner-savior he can really do no wrong because he is sent to Earth on a mission. Aware of this, Trump himself claimed during the presidential campaign that he "could stand in the middle of Fifth Avenue and shoot somebody and I wouldn't lose voters."[36] There is also little doubt that his fervent believers could easily become part of a real physical civil war (not just a cold one) if he were deposed from power and incited to take up the sword in a crusade to protect him.

As an aside, it should be emphasized that not all religious groups in America think in this way. Although they may support Trump's "morality-restoring" agenda, they certainly do not see him as a savior, nor accept his racist narratives, profane speech, personal immorality, and overall bluster. So, it is not religious people per se who support Trump, but those who espouse an apocalyptic belief system that he can easily manipulate. Mussolini too espoused moral causes opportunistically, closing down wine shops and nightclubs, which were seen by religious people of the era as signs of degeneracy, perversion, and sinfulness. He also made uttering profane and obscene language in public a crime, and he pushed the view that women should stay at home and look after their families while their husbands worked—a model of family life endorsed enthusiastically by the Church. He also opposed the use of contraception and divorce. The parallels with Trump are remarkable, given the latter's stances on abortion, on women's role in the world, and similar Mussolini-type positions that are seen by many religious groups as critical to the restoration of moral order to American society.

Regardless of empirical evidence that Trump is a fake believer who uses religion as Mussolini did for self-serving purposes, he is still seen conveniently as a culture hero—a topic that will be examined more closely in subsequent chapters. Psychologist Frederik Lund has provided an explanation of why belief is so unshakable. The abstract from his study is worth reiterating here in its entirety:[37]

> Belief has a large emotional content. On the basis of the results on the rating of a series of propositions on a belief scale by college students, the correlation obtained between belief and desire was +.88. This confirms in some respects the theories of the psychoanalytic school and the dynamic psychologists in the evidence for an emotional and instinctive basis for motivation and belief. It was also found that there is a marked tendency to idealize the rational principle and to conceive of it as the most valid and important of belief determinants, notwithstanding the fact that non-rational factors appear to outweigh it so largely in conditioning our belief-attitudes. The fact that beliefs once formed are not willingly relinquished is definitely related to, if not responsible for the fact, that the side of the question first presented to us, and the first influences brought to bear upon us, are most effective in determining our beliefs, so much so as to suggest the presence of a law of primacy in persuasion. Belief, as a certain mental content, is present throughout the scale of knowledge and opinion, just as is temperature on a scale the extremes of which are hot and cold; it is not present with the same strength, however, but with varying admixtures of doubt.

The ancient Greek philosophers divided belief into *pistis* and *doxa*. The former implies a sense of trust in something, and the latter a system of opinions that guide actions and behaviors no matter what the truth of the matter. They coalesce to produce the tendency in all of us to view reality in a binary way. In belief, there are only binary choices—something is either true or false, right or wrong, moral or immoral, and so on. In a series of essays called *Illustrations of the Logic of Science*, the American pragmatist philosopher Charles S. Peirce described belief as something that impels us to act, not just a state of mind.[38] He defined it as a habit or rule of action that we easily adopt to counteract doubt (the opposite of belief), from which we struggle to free ourselves. In other words, belief is the emotional strategy we use to eliminate the burden of doubt, and language is the vehicle used for establishing and reinforcing beliefs. In fact, if we listen carefully to believers, such as the women interviewed by CNN above, we can literally hear their beliefs coming out in what they say.

EPILOGUE

As Martin Luther King Jr. once put it, in response to the lies that were being hurled at him, "No lie can live forever."[39] Dictatorships come and go, confirming the veracity of Dr. King's aphorism. The falsehoods on which Soviet communism, fascism, and nazism were based were eventually exposed by the force of truth and objective facts. The implication that can be gleaned from King's statement is that the lies that Machiavellian liars perpetrate for opportunistic reasons will eventually dissipate. True political discourse involves making rational arguments and using evidence in support of these (*lógos*), not on beliefs and opinions alone (*mythos*). To return to Senator Patrick Moynihan's admonition, used as the epigraph to this chapter: "Everyone is entitled to his own opinion, but not to his own facts."[40]

The Russian hacking of the 2016 American presidential election campaign put on display how easily people can be duped by doublespeak and by the strategy of alternative facts.[41] The use of conspiracy narratives with regard to race, the economy, law and order, immigration, and liberalism constituted the mythological subtext that became, cumulatively, a subtle rallying cry for breaking the norms of American democracy through the leadership of Donald Trump. However, as Dr. King eloquently stated, truth is a powerful antidote to mendacity, even though it takes time for it to foment in people's minds.[42] One of the most famous of Aesop's fables, which is based on this principle, is that of "The Boy Who Cried Wolf." The story tells of a child who continually "cried wolf" falsely in order to draw attention to himself, until one day he was actually threatened by a real wolf. When he cried wolf again, no one believed him, thinking that it was just another lie.

Mussolini's fall from grace started when the radio news media turned against his alliance with Nazi Germany and his policies of retrenchment that were taking their toll on Italy's economy and spiritual well-being. The Italian media had seen the "writing on the wall"—an expression that goes back to the Book of Daniel in the Bible when, during a feast for King Belshazzar, someone mysteriously appears to write the Aramaic words *Mene, Mene, Tekel, Upsharin* on a wall, which translates as "Numbered, numbered, weighed, divided." Daniel interprets this as a warning pointing to the downfall of the Babylonian Empire. The English phrase *our days are numbered* is a derivative of this ominous warning. The moral is that the Machiavellian liar will eventually have to confront the writing on the wall, on which his days are said to be numbered.

Belief systems are not monolithic; that is, we may entertain beliefs that may even contradict one another at different levels and in different situations. The mind is a marvelous "blending" organ that allows contrasts to coexist, compartmentalizing them in such a way that we can go from one to the other in specific situations with cognitive ease and with no dissonance whatsoever. The master liar will know how to break this pattern of blending by creating confusion, whereby nothing can be believed. This is what characterizes doublethink, as Orwell insightfully pointed out and cited above but worth repeating here for emphasis:[43]

> To know and not to know, to be conscious of complete truthfulness while telling carefully constructed lies, to hold simultaneously two opinions which cancelled out, knowing them to be contradictory and believing in both of them to forget whatever it was necessary to forget, then to draw it back into memory again the moment it was needed, and then promptly to forget it again. That was the ultimate subtlety: consciously to induce unconsciousness, and then to become unconscious of the act of hypnosis you had just performed. Even to understand the word "doublethink" involved the use of doublethink.

The ability to hold contradictory thoughts in the mind is actually an extremely useful one, constituting, as F. Scott Fitzgerald observed, a unique form of intelligence: "The test of a first-rate intelligence is the ability to hold two opposing ideas in mind at the same time and still retain the ability to function."[44] The Socratic dialogue is based on contradiction. It aims at first to evoke disagreement on some belief, so that such disagreement can be resolved logically by the interlocutor. The typical resolution path sees Socrates inveigling his opponent to consider certain other beliefs until a contradiction is reached by implication. Doublethink is, in contrast, a state of mind in which there are no hypotheses or counterarguments; only manipulated beliefs. As intimated in Orwell's quote above, doublethink relies on confusion, a kind of imposed circular reasoning that leads nowhere and that can thus be easily constrained, contained, and controlled.

The psychological reason why lies will eventually come tumbling down, as Dr. King suggests, is because contradictory beliefs cannot continue to coexist in the same brain unless they are resolved—a process called the resolution of cognitive dissonance by psychologists, which will be discussed in the final chapter. Belief is, in fact, what alternative facts, conspiracy narratives, and the like are designed to create. Without it, the wall of lies will come crumbling down by itself.

3
CONFABULATION

Those who do not remember the past are condemned to repeat it.

—George Santayana (1863–1952)

PROLOGUE

We all tell white lies about ourselves to others and even to ourselves—in order to embellish our life stories, to make them fit in with the situation in which the telling occurs, to brag a little, and so on. This kind of "autobiographical white lying" is accepted tacitly by everyone as part of the game of social interaction, and we make very little of it, knowing that there is always some element of truth in the story. Eventually, as we also suspect, most of the truth will come out through subsequent contacts, conversations, experiences, and interactions. This type of self-serving autobiographical tale is designed not to mirror what may actually have happened in one's past, but mainly to explore and interpret one's personal past for an audience. In cyberspace, this type of tale, based partly on truth and partly on fabrication, is found throughout social media platforms. In a nutshell, the persona we present to others is largely "confabulated"; that is, we have woven it together through recollection, partial fabrication, and some self-aggrandizement. Confabulation can be defined for the present purposes as an account based on an interpretation and recreation of the past, via subtle alterations and even fabrications.

Everyday confabulations are essentially harmless and, since everyone tends to spin them, they are perceived as part of the construction of the "ideal self," as some psychologists have claimed.[1] But confabulation does not end at the level of the individual; it is typically extended as an account of the past that is either completely made up (false) or based on bits and pieces of truth stitched narratively together in such a way as to present the past in some self-serving way. In the hands of the cunning liar-prince, confabulation allows him to manipulate people's perceptions of the past and direct them toward his ultimate goals. Confabulated autobiographical histories are told by pathological liars, con men, hucksters, criminals, and liar-princes alike—only the details of the narrative change according to situations and needs. The objective of this chapter is to look at the nature of false accounts of history to incite people to act in the service of some illusory ideal or to support the liar-prince. The term *confabulation* will be used exclusively with this definition throughout the chapter. It is yet another one of the shrewd tactics in the Art of the Lie. The liar-prince is, thus, not only a clever wordsmith, but also a master storyteller, who presents himself as an ersatz "wise elder" that people should trust and whose version of history is the only plausible one.

Confabulation comes in two forms—partial or total. The former involves the incorporation of actual events that are told in traditional histories into the confabulated narrative, molding the real and the fictional into a storyline that taps into inherent beliefs about the past. Over time, this type of confabulated story starts to take on higher and higher degrees of verisimilitude, making it difficult to dispel it with counterarguments and contrastive empirical evidence. The latter type of confabulation inheres in the total fabrication of the past, constituting a pure form of mythology. In politics, this kind of confabulation can have monumentally dire consequences. Some of the most heinous confabulations have, in fact, wreaked havoc upon humanity. One of these was Adolf Hitler's mythology of a "master race," which he grafted from a false narrative based on a purported ancient and proud "Aryan race." This was a pseudo-classification of Caucasian (white) people that surfaced in the latter part of the nineteenth century to exalt such people as crucial to the progress of humanity. It was obvious from the outset that this was a confabulation, as the linguist Max Müller wrote in 1888, stating that anyone who "speaks of Aryan race, Aryan blood, Aryan eyes and hair, is a great sinner as a linguist."[2] Hitler adopted the Aryan myth nonetheless to perpetrate his imperialistic anti-Semitic and white supremacy bigotry, at the same time that he could use it as justification for world domination by a "master race." Mussolini adopted Hitler's Aryan myth, as he made clear in a speech he gave in Bologna in 1921, claiming that fascism was born "out of a profound, perennial need of this our Aryan and Mediterranean race."[3] However, he ultimately rejected the notion of a biologically pure race, at least in his private statements.

In a phrase, when deployed by liar-princes confabulation can have deleterious human consequences. Now, the same kind of Aryan myth has reared its ugly head in many parts of the world today, including the United States, where it is used to promote "white supremacy" and "neo-Nazi" sentiments and movements. Donald Trump has been ambiguous with regard to this type of mythic confabulation, straddling the line precariously between feigning ignorance of the mythology and engaging in perilous forms of tacit acceptance. His MAGA moniker falls into the domain of confabulation, subtly incorporating suggestive elements of a "pure (white) race" narrative that founded America and which, because of liberalism, has been marginalized or destroyed. MAGA is the "cover-page title," so to speak, of a confabulated story of White America. Its central objective is to bind people of like mind together through a false narrative that is self-serving and highly manipulative of the historical facts. Confabulation of this kind is effective because it works at a subconscious level, stoking suppressed dormant biases that had been purportedly relegated to the dustbin of history by the deep state.

THE NATURE OF CONFABULATION

We are a historical species; that is to say, we evolve not only biologically but also through culture. We record our cultural evolution through narratives that create a sense of continuity from one era to the next and through which we define ourselves. Historical accounts start, as discussed in the previous chapter, with foundation myths that are told to explain how we came into existence. Later eras produce stories of heroes and their legendary exploits. Examples include the story of Robin Hood in England, who stole from the rich in order to give to the poor; William Tell in Switzerland, who resisted tyranny and played a dominant role in Switzerland's liberation from Austria; Davy Crockett in America, who died bravely in the Battle of the Alamo in 1836 for Texan independence from Mexico. Britain, Switzerland, and the United States have made these heroes, and their stories, part of the cultural fabric of their societies. The kind of history we read in school, which connects dates, events, and personages into an overarching chronology, is, more technically, a historiography—an archival narrative that provides a mnemonic memoir of a society's past and its relevance to the present along with its implications for the future.

Histories are never completely true. They are interpretations and are thus analogous to the white autobiographical lies we tell about ourselves—partly true and partly altered to impart a sense of coherence and meaning to the plot. They are interpretations by historians or storytellers. This crystallizes not in the actual "chronological facts" themselves but in the way they are put together, in much the same way that the interpretation of sentences in a language is not based on the meanings of the individual words that constitute it, but on the syntax and semantics that connects them holistically. Psychologically, we seem to need to connect the dots in our own lives through the story format. Similarly, societies need to connect historical dots in order to establish a continuous record of important events that have made them distinct and meaningful to those reared in them.

When groups or communities are established for some particular reason, such as the Mafia, falling outside the paradigm of traditional historiography, they seek legitimacy by fabricating their origins and historical evolution—hence their recourse to the strategy of confabulation. Criminal societies make up stories about their origins in order to gain historical justification, otherwise they would be seen as mere thugs.[4] Like the founding myths of ancient societies, the confabulated histories invariably trace the criminal gang's origins to some meaningful event, which imparts an aura of legitimacy that ensures that gang members, extant and new, believe that their raison d'être is justifiable and authentic. This imparts a sense to gang members that they have a right to do what they do because it is their destiny. In effect, by spinning an unfounded story about its origins, the Mafia aims to cast itself into the domain of legitimacy as a recognized organization keeping members united with a sense of purpose and continuity.[5] In reality, organized criminal gangs in Italy's south were born from its exploitive past feudal system. So, the confabulated history takes the factual materials from this past and then reorganizes them into a collage of events that impart a sense of purpose to the foundations of the criminal gang.

To see what this implies, it is worthwhile to digress momentarily to examine how the confabulated Mafia story originated and evolved.[6] The absentee noblemen of Sicily's feudal system needed strong men with local power and influence to manage their estates while they were away, and who would be willing to use strong-arm tactics to ensure that the feudal residents did not vandalize their estates or engage in acts of robbery. Ruffians were hired as both personal guardians and estate sheriffs. As feudalism receded, the enforcers did not go away, becoming independently powerful simply by staying together as a group. Given that government officials were negligent and legal proceedings were lax, byzantine, and corrupt, these thug-guardians gained the trust of common folk, even though they exploited them through blackmail and extortion. Corruption was rampant—judges bought their posts; lawyers were paid little or nothing, negotiating fees feloniously; police officers were often unscrupulous and untrustworthy. Having no faith in the authorities, common people looked to the guardians for protection. Banditry thus established itself as a protection business. By the nineteenth century, it spread throughout Sicilian society. During the 1848 rebellion against Sicily's Bourbon rulers, the bandit-guardians joined the uprising, allying themselves with Giuseppe Garibaldi, the patriot and military leader of the unification of Italy in 1860. After unification, they gained even more legitimacy and power, as the authorities once again fell prey to corruption. As crime historian Paul Lunde notes, it was the "traditional Sicilian suspicion of state institutions that created the conditions in which the Mafia could develop."[7] Nothing has changed since then.

Clearly, this type of historical background does not benefit the Mafia. So, in one of its confabulations it traces its origins to two medieval codices, *Breve Cronaca di un anonimo cassinese* ("History of an Anonymous Cassinese," 1185) and *Cronaca di Fossa Nova* ("History of Fossa Nova," 1186), which describe a secret organization of *Vendicosi* ("revenge seekers"), whose members were punished severely, and even hanged, by the king of Sicily for various crimes they were said to have committed. But the real reason for the attempt to eradicate the organization was that it threatened the power of the king, who realized that it was a "confraternity" of street gangsters who could be hired by adversarial or antagonistic noblemen to do their bidding against him. In 1784, while visiting Sicily, author Federico Münter wrote about one of these confraternities—known as the confraternity of Saint Paul—which was founded in the sixteenth century, during the reign of Charles V. The members claimed to protect orphans and the oppressed, although the reality was rather different—they offered protection to anyone for a price. From this story the legend of the *Beati Paoli* emerged, which are portrayed by some Mafia accounts as the predecessors of the criminal gang. The legend became, in effect, a convenient confabulation, since it allowed the Mafia to portray itself in a positive light as an organization whose primary mission was to fight oppression. Even a novel, published in 1909 to 1910 in serial form in a daily newspaper of Palermo, picks up on this legend and enshrines it into popular lore even more deeply. Written by Luigi Natoli, the novel became a reference point for justifying the origins of the Mafia.[8] In it, the nobleman protagonist, Coriolano della Floresta, creates an alternative justice system, to which those who have been oppressed and who distrust the authorities can resort.[9]

The point of the foregoing discussion is that criminal gangs make up their origins, portraying themselves as descendants of chivalrous confraternities or brotherhoods, and thus differentiating themselves from common, everyday street hoodlums. These fabricated legends are designed to portray the gang members as folk heroes, as criminal analyst John Reynolds observes:[10]

> Like Sherwood Forest outlaws, the Sicilian bandits created their own folk heroes, lauding their bravery and exploits as examples of gallantry. The most celebrated of them, a man named Saponara, was captured and imprisoned in 1578. According to Sicilian lore, Saponara was tortured by his Spanish captors in an effort to learn the names of his cohorts but Saponara chose to die in agony rather than betray others. His bravery became a symbol for every Sicilian who believed their salvation could be achieved only through loyalty.

Confabulated stories have an enduring impact on the minds of both insiders (the gang members themselves) and outsiders who may sense some truth in the fabrications simply because they are grafted from events that also inform real histories. They become unconscious myths. As the late French anthropologist Claude Lévi-Strauss observed: "Myths operate in men's minds without their being aware of the fact."[11]

Hitler's Aryan myth is a horrific example of what confabulation can instill into groupthink—a hatred of outsiders who are seen as upsetting the historical destiny of the Aryan people that its forefathers had envisioned. Eradicating the "others" through any means possible became the rallying cry of the Nazis, leading to horrendous events such as the Holocaust. Once followers inserted themselves into the apocryphal storyline, they saw themselves as valiant soldiers in the battle for racial hegemony. This makes it virtually impossible to cast doubt on the story's validity, given the high degree of emotional commitment made to it by individuals. Once drawn into it, escape from it is virtually impossible for the simple reason that no one wants to admit to having been duped or defrauded by the confabulation. This is the same kind of reaction of those who have been duped or swindled out of money by con artists. Rather than have to face the truth and admit that they have lost everything by believing a swindler, it is much easier for them to go into denial, blocking the truth from becoming part of their conscience as a defense mechanism. Confabulation allows the Machiavellian liar to take hold of people's beliefs and twist them for his own objectives. As a result, the liar is not seen as a con artist or deceiver, but as a possessor of the hidden truth who fleshes it out by identifying the villains in the story.

Confabulated histories are not exclusive to dictators and criminal gangs—they can be spun by anyone. It is chilling to consider, for example, that a pseudo-foundation myth of America was projected (literally) onto the early cinema screen—namely, the myth of white supremacy evident in the immensely successful 1915 silent film *The Birth of a Nation*, directed by D. W. Griffith, and adapted from the novel, *The Clansman*, by Thomas Dixon Jr. (1905).[12] The movie was brazenly racist, even though Griffith apologized, after strong criticism, for the adverse effects the film had brought about. The plot revolved around the supposed key role that the Ku Klux Klan played in the origins of America, implicitly suggesting that the Klan was a founder of America and thus that the nation was built on the cultural heritage of the white settlers. The movie is one of the first successes in early cinematic history, rousing controversy to this day. To be fair, there is no biographical evidence to suggest that Griffith himself was a racist. He claimed to have incorporated the story of the KKK in order to portray American history realistically, rather than idealistically. His subsequent film, *Intolerance* of 1916,[13] was a lengthy epic covering four historical periods. It is seen as his apology for *Birth of a Nation* and a portrayal of the horrible effects of human cruelty on the progress of civilization.

Confabulations such as *The Birth of a Nation* are based on events that may or may not have actually occurred in the way they are presented, but they are stitched together in such a way that they fuel believability through verisimilitude. In other words, they are presented as being based on true events, but are in fact confabulations. As social critic W. T. Anderson has observed, such representations are perceived as plausible because they fit in with extant belief systems and worldviews.[14] The boundaries between the imaginary and the real break down in confabulated narratives, even those that are completely false, because there is an unconscious desire for them to be true in some people. A made-up story such as the Aryan myth gains believability gradually and broadly as it spreads through all kinds of representations, such as speeches, newspaper articles, radio, film, and more. This is what happened in Nazi Germany. As the French writer Marcel Proust so aptly put it, "Time passes, and little by little everything that we have spoken in falsehood becomes true."[15]

Confabulated histories are everywhere today, especially in cyberspace, where they compete with "official" histories. The truth is, literally, what the confabulated story says it is. To cite Anderson, the confabulators "take the raw material of experience and fashion it into stories; they retell the stories to us, and we call them reality."[16] The late French social critic Jean Baudrillard maintained that the borderline between fiction and reality has utterly vanished in modernity because confabulation has become an unconscious language, inducing a mindset he named the "simulacrum," whereby what occurs on the screen and what occurs outside of it in real life are perceived as reflexes of each other.[17] This leads to a disruption of the normal functioning of the brain's perception mechanisms, which are designed to differentiate between imaginary and real events. The term *confabulation*, actually, comes from clinical psychology referring to a disturbance that manifests itself in distorted or misinterpreted memories about the world. Confabulations are simulacra that twist people's perceptions of reality. In them, there are heroes and villains, conflicts and victories, and successes and defeats. It is this aspect that makes them particularly perilous, because the villains are those identified by the confabulators themselves. In the case of myths of racial purity, the villains are those who are not part of the master race.

The narrative in *The Birth of a Nation* conveniently ignores the struggle for racial equality that has been playing out in America since the days of slavery. A more honest history of America would validate the many contributions of African Americans to the nation. In any utilization of the Art of the Lie, there is perhaps no more dangerous strategy than this kind of confabulation, since it stokes hatred of otherness, which is seen as a threat to the supposed cultural hegemony of a nation.

The main antidotes to the destructive effects of such mythic narratives include social activism, such as civil rights movements, and counternarratives, such as those that provide a more accurate understanding of the role of racial diversity and the deleterious effects of racism in America. In the latter category are movies such as *Mississippi Burning* (1988),[18] *Ghosts of Mississippi* (1996),[19] *A Time to Kill* (1996),[20] and *BlacKkKlansman* (2018).[21] Fascism emerged in the 1920s, the decade when cinema became a powerful new medium of artistic and political expression. Fascist regimes were keen on tapping into the emotional power of cinema to excite their audiences, at the same time that cineastes used the medium as a means to oppose fascism and extreme nationalism. The new medium thus became ipso facto both the voice of the resistance to fascism and a means for fascism to portray itself in a positive light. Today, cinema continues to be one of the most effective voices of the resistance to all forms of fascism. This is the reason why it is typically censored in totalitarian states.

REDEMPTIVE HISTORY

In a 2018 January interview with the Christian Broadcast Network, Trump's press secretary at the time, Sarah Sanders, made the surreal claim that Trump was sent to Earth by God, supposedly to enact the conservative evangelical cultural agenda that would restore Christian biblical beliefs as the basis of morality in America—an agenda that had been purportedly disrupted by the forces of rampant atheistic liberalism and relativism associated with previous presidencies.[22] For individuals such as Sanders, the MAGA narrative is perceived to be a redemptive one, implying a restoration of America's Christian heritage, in opposition to the view of America as a culturally diverse society (religiously, ideologically, and ethnically). This not-so-subtle interpretation has, needless to say, racial overtones, since the original mythic heroes of the narrative are the founding group of white colonists who had, themselves, escaped religious persecution in England. The MAGA narrative for white evangelical Christians is a new chapter in America's true history, aiming to restore the religious values that they see as foundational to America. The narrative also incorporates the theme of the American Dream, whereby success can be achieved through thrift, hard work, and frugality. But that dream, for the slaves, was and may continue to be a nightmare. The MAGA narrative is thus an attempt to regain and control the past in an exclusionary, rather than inclusionary, way. As Orwell wrote in *1984*, "Who controls the past controls the future: who controls the present controls the past."[23]

MAGA excludes not only the role of slavery in the foundation of America but also the role of indigenous societies who were in America before the arrival of the colonists. It also conveniently ignores the critical role of immigration to America's sociocultural evolution—a fact emblemized by the Statue of Liberty. MAGA is a shrewd, crafty confabulation, which aims to redeem a pristine past that was hardly that. As such, it has provided Trump and his acolytes with a powerful confabulation tool that resonates emotionally with those who believe that they have been marginalized or rebuked by those who promote diversity as a political tactic. It is, in this sense, a purification story.

The Pilgrims and the Puritans established the first English-speaking communities in America based on the religious traditions, practices, symbols, and rituals that they brought with them from England. The settlers held a fervent belief in the socially binding functions of religious celebrations, generally revolving around elaborate meals at Easter and Christmas as well as religious songs and dances performed ritualistically at specific periods of time. All frivolous entertainments were strictly prohibited. As followers of Oliver Cromwell in England, the Puritans in particular frowned upon any libertine, licentious, or gluttonous lifestyle, seeing it as sinful, degenerate, and leading to eternal damnation. They insisted on sobriety, plainness of dress, and rigid moralism within their communities. The colonists also established new religious festivals as part of their adaptation to the new environment. Thanksgiving, for example, originated as a harvest festival in the late 1700s to give thanks to God for the plenitude of the land. As Abraham Lincoln later declared in 1863, Thanksgiving was about giving "praise to our beneficent Father who dwelleth in the Heavens."[24] The first Thanksgiving was celebrated by the Pilgrims in 1621, introducing the tradition of consuming a turkey meal as symbolic of the abundance and well-being that the harvest brought about. To this day, turkey has a higher symbolic value than any other kind of meat dish in America, a fact that traces its roots to the original Thanksgiving feast. Of course, the unconscious sense of sacredness associated with a turkey meal is a contextualized one; that is, its historical symbolism is relevant at a Thanksgiving meal, not in the eating of a common, everyday turkey sandwich.

The colonist lifestyle, based on frugality, diligence, temperance, and industriousness, has been designated the "Protestant work ethic," considered to be the founding *esprit* of American society. Over time, the colonists became wealthy through thrift and hard work, eventually taking over the reign of economic power in the bustling cities that started springing up in the latter part of the nineteenth century. The origins of modern corporate capitalism are to be found in this ethic, as sociologist Max Weber cogently argued at the turn of the twentieth century.[25] Weber suggested that the religious lifestyles and values of founding groups such as the Pilgrims and the Puritans unconsciously created a version of capitalism that sees profit as a virtuous end in itself, and thus as a goal that must be pursued as if it were the equivalent of religious virtue. Weber claimed that this transformed the traditional forms of capitalism in Europe, based on the ownership of businesses and economic enterprises by families, and paved the way for contemporary corporate capitalism. Once this had emerged, the original Protestant religious values were no longer required to be espoused in an overt public way, given that the ethic on which they were based took on a new economic form. It is within this social environment that the so-called American Dream emerged, whereby material success is afforded to anyone who subscribes to the work ethic. If one does not, then one is ipso facto seen as antithetical to "real" American values. In this historical paradigm, slaves and indigenous people are often seen as outsiders if they do not subscribe to the same ethic. Only when they do are they likely to be accepted within the paradigm.

As Arthur Asa Berger has cogently argued, we hardly realize the extent to which America was founded by, and then shaped in reaction to, the Protestant ethic. America's character developed when, as Berger puts it, the early colonists decided that consumption had a "place in God's scheme of things."[26] Consumption was thus rationalized as an earthly reward for diligence and hard work. As Berger goes on to observe, "There is, indeed, an important religious or sacred dimension to our consuming passions."[27]

The Protestant ethic started to weaken in the latter part of the nineteenth century, when the shift toward a more secular America started fomenting in the new urban centers. Among the first to pave the way to "hell on earth," as the moralists of the era described it, was a New York

City showman and circus operator—P. T. Barnum. Barnum's spectacles were the antithesis of Puritanical restraint, providing sinful pleasure and delights to anyone, regardless of their social backgrounds. They were egalitarian by happenstance—that is, they were blind to people's race, ethnicity, class, and even tastes, since money knows no discrimination. At the "Greatest Show on Earth," as Barnum called his circus, which he founded in 1871, everyone was welcome, regardless of race or ethnicity. Moreover, the spectacle itself was indirectly subversive to Puritanical values, since anyone could find something profane to amuse them, and none of the performances, especially the prurient sideshows, had anything to do with religious values or mores. Barnum's influence on America's shift away from Puritanism to secularism, and his opening the doors, literally, to the participation of "others" into the entertainment mix, cannot be overemphasized. If one were to locate a point in history when the breaking of the Puritanical umbilical cord occurred, and when consciousness of diversity emerged, it is likely to be in the entertainment spectacles that circus culture, spearheaded by Barnum, introduced into America. The circus was, needless to say, condemned by many at first as sinful, not only because of its profane spectacles, but also tacitly because it allowed the heathen "others" to be a part of the show. The tide could not be turned, since a new esprit de corps was simmering in America that eventually would lead to liberation from the strictures of Puritanism. The circus was, figuratively speaking, a tipping point for the crystallization of a new sense of a diversifying America, legitimizing a place in society for everyone, no matter their race or ethnic background. The circus took America more and more away from its white Protestant roots. It did so not because of any shift in philosophical worldview, but because of the profit motive. Ironically, this led to a society that was becoming more and more open to diversity.

As affluence spread in the new burgeoning cities at the start of the twentieth century, Americans started to have more leisure time at their disposal. Social life came to be characterized by workweeks and weekends—the latter becoming increasingly marked by a surge in recreational activities, many of which involved the participation of the "others," including African American jazz musicians. Jazz was initially deemed to be obscene and vulgar by many social elders, primarily because it was the music of the "slaves." However, in true entrepreneurial style, and much to the chagrin of the moral guardians of the era, jazz and its attendant lifestyle spread broadly among young white people. New profit-seeking entrepreneurs came forth to provide outlets for the new forms of entertainment based on African American culture to thrive, leading eventually to the Roaring Twenties and to the establishment of an ever-expanding culture of diversity.

The 1920s saw unprecedented economic growth, rising prosperity, and far-reaching social change that involved an ever-increasing acceptance of racial and ethnic otherness, as well as immigrants; but the era also saw the rise of fascism in Europe, followed by nazism in Germany and, in America, the rise of white supremacy groups such as the Ku Klux Klan. After World War I, America started to diversify its social make-up even more through an increasingly open immigration policy. Throughout the twentieth century, the nation became a true melting pot, promoting ethnic, religious, and racial diversity as part of its ongoing experiment in diversity. This came to an unexpected head with the election of an African American president, Barack Obama, in 2009. In the first decade of the twenty-first century, America had seemingly become color blind and welcoming of people of any race or background within the halls of political power. The American Dream was finally being realized by truly anyone. But old habits and beliefs die hard, as the saying goes. To some, the election of a Black president was seen as catastrophic, leading to a form of reverse discrimination, with white culture being relegated to the margins and even denigrated under his presidency. The seed of a cultural civil war were sown by the end of Obama's tenure as president, opening up the way for MAGA to imprint itself into the mindset of many who resented Obama's presidency.

Aware of this brooding resentment, Donald Trump emerged literally from the shadows of the past to promise a redemption and retrieval of America's "real" past, even designating himself as the one who alone could redeem the real past, imprinted in his Twitter handle, "@realDonaldTrump." The MAGA narrative, which became the central confabulation of Trumpian politics, attracted those who felt marginalized by the Obama presidency, including fervent white evangelical groups. Restoring the founding religious culture to America became tacitly embedded in the subtext of the MAGA narrative. Restoring morality, eliminating relativism, and defeating the chaos of diversity were the main elements of the MAGA narrative. Many evangelicals came to see Trump as the spiritual vessel through which such historical redemption could be realized. Through favorable radical conservative policies, judiciary appointments, and the placing of key members of the religious community to positions of governmental power and influence, Trump has emerged as the champion of redemption politics.

As any strategy in the Art of the Lie, the MAGA narrative depends on Orwellian ambiguity for its effectiveness. To those who see through Trump's confabulation technique, the MAGA slogan is perceived to be racist code, but to evangelicals like Sarah Sanders, it is perceived instead as redemptive and exculpatory. The sense that America had been overtaken by secularists was a real one during the 2016 presidential campaign, and it was this fear that the MAGA narrative exploited among the evangelicals.

As a skilled confabulator, Donald Trump is well aware of the emotionally charged ambiguity of his narrative. It keeps his followers solidly behind him, since they are not protecting him as an individual, but for what he stands for. Trump can thus lie constantly, knowing that he will still maintain a solid base of followers, simply because they are committed to the redemptive politics that he promised to instill into the fabric of government. Trump knows, in other words, that his MAGA followers are willing to be deceived because there is a greater moral cause at stake.

As Machiavelli observed, this is one of the most powerful weapons at the disposal of the liar-prince:[28]

> It is necessary to know how to conceal this characteristic well, and to be a great pretender and dissembler. Men are so simple, and so subject to prone to be won over by necessities, that a deceiver will always find someone who is willing to be deceived.

The liar-prince is a master magician of the art of confabulation, able to adapt his false narratives to audiences through prevarication, knowing full well that his followers are easily duped by it. In the *Tao Te Ching*, a philosophical treatise attributed to Lao Tzu, the sixth century BCE Chinese philosopher, it is written that "Everywhere men yearn to be misled by magicians."[29] The MAGA narrative is clever sleight of hand, since it is open to varying interpretations, from religious to political ones. The ploy, as magicians certainly know, is never to make any serious mistakes in public when telling the story. The stratagem is to pretend to be what followers want the story to be about, as Machiavelli points out: "A prince ought to take care that he never lets anything slip from his lips that is not replete with the qualities, that he may appear to everyone who sees and hears him as a paragon."[30] In other words, the confabulator must never be caught open-handed, because, if he is, he will not be able to survive the exposure. He must always be a master of deception and distraction, keeping both followers and enemies in a state of confusion through the fabrication itself. As linguist Frank Nuessel observes, this is an inherent principle of confabulation used by military and political leaders to great effect.[31]

OTHERNESS

The late French social philosopher Michel Foucault saw, as one of the primary tactics of political repression, the attack on "otherness"—that is, on those who are not perceived to fit the racial or ethnic profile of a society and, thus, assumed to be a threat to the homogeneity and hegemony of the dominant group.[32] Attacking otherness has always been a goal of sinister political confabulation, from the Aryan myth in Germany to current racially based conspiracies that go viral through social media depicting "the others" as destroying America. Outsiders are blamed for causing society's troubles and are thus vilified as immoral invaders or intruders. For this reason they must be thrown out or blocked from coming into the society. Trump's wall metaphor is intended not only to keep illegal immigrants out physically, but also symbolically.

Confabulated histories such as the Aryan myth are, as mentioned, fabricated to imply that the hegemony of the founding tribe is under attack by those of different racial backgrounds. The MAGA story is a similar attack on otherness. This does not necessarily imply that believers in the story are racist. The subtlety of the narrative is that it talks indirectly to all kinds of people, from religious moralists to those who feel left out of the mainstream. It is an Orwellian strategy, crafted to restore pride in the supposed historical roots of the "Real America," and thus to restore its "real culture." In the process it attacks otherness as a source of the disruption of these roots.

It is useful to consider Hitler's Aryan myth a little more closely so as to be better able to grasp why mythic confabulations are effective. The term *Aryan*, as indicated at the start of this chapter, was coined in the nineteenth century in reference to a family of ancient languages spoken by people in the Indian subcontinent, whose origins are traced as far back as 1500 BCE. The term was initially just that—a classificatory term used by language scientists to identify a specific group of languages. However, it started to resonate with false notions of a master race already in that era, migrating beyond the realm of linguistic classification to the domain of prejudice and bigotry. It did not, however, gain a foothold broadly anywhere, until the Nazi Party came to power in Germany, led by Hitler, who twisted the meaning of the word *Aryan*, narrowing it to refer to a pure race destined to rule the world. In his delusional mind, that race was constituted by the Germans themselves and a few other northern Caucasian peoples. This false use of the term, alluding to a pure white race, has continued among supremacist groups and neo-Nazis throughout the world. The Aryan myth was designed to identify a single race as the "chosen" one (by biology and history) to lead the world; relegating others as inferior and, more importantly, as interfering with the destiny of the chosen race. In Nazism the "others" were initially Jews, Slavs, and other minority groups. The Aryan myth emerged to justify the elimination of such groups, and even extermination (as in the Holocaust), so that the master race could finally rule the world unimpeded and build a harmonious, orderly, and prosperous civilization. With economic and other social problems rampant in Germany, the Aryan myth came forward to propose a reason why such problems existed—they were caused by the inferior groups of people. Many in Germany believed this myth, accepting its nefarious premise of a master race that would bring about peace, social harmony, and progress. But the truth of the matter turned out to be just the opposite. Hitler's regime brought about terrorism, war, and the Holocaust instead of harmony and prosperity.

Belief in racial superiority is not exclusive to a particular society or a specific era. It has existed since the dawn of history. Ironically, the ancient Romans saw the Germanic tribes as a race of barbarians that was barely human. The American colonists claimed superiority over the Native American tribes, justifying the appropriation of their lands. Around ten thousand years ago, the members of individual tribes sought larger territories with more natural resources within which to live. This led to what the anthropologist Desmond Morris calls the formation of supertribes—expanded groupings of people that came about as a consequence of tribal expansion and tribal admixture.[33] The first supertribes date back around five thousand to six thousand years, when the first civilizations came onto the scene, defined anthropologically as collectivities of individuals who, although they may not all have had the same tribal origins, nevertheless participated, by and large, in the culture of the founding or conquering tribe (or tribes). Unlike previous tribes, supertribes enfolded more than one cultural worldview. So, unlike what the racial purists believe, since the dawn of civilization, people of different races (tribes) had to learn how to live together and cooperate in order to survive. Nevertheless, we classify and think of ourselves as members of distinct races and ethnic groups—that is, as belonging to groups of people with common genetic links, even though admixture has always been the pattern in the foundations of societies. No two human beings, not even twins, are identical. The proportions of traits, and even the kinds of traits, are distributed differently from one part of the world to another. As it turns out, these proportions are quantitatively negligible. Geneticists have yet to turn up a single group of people that can be distinguished by their chromosomes. There is no genetic test or criterion that can be used to determine if one is racially or ethnically, say, Caucasian, Slavic, or Hopi. Populations are constantly in genetic contact with another. The many varieties of *Homo sapiens* belong to one interbreeding species, with little genetic difference among individuals. In fact, it has been established that 99.9 percent of DNA sequences are common to all humans.[34]

So, from a purely biological standpoint, human beings defy classification into races or ethnic groups. Nevertheless, the historical record shows that from ancient times people have, for some reason or other, always felt it necessary to classify themselves in terms of such categories. The Egyptians, the ancient Greeks of Homer's time, and the Greeks and Romans of classical times, for instance, left paintings and sculptures showing human beings with perceived racial differences. And most languages of the world have words referring to people in terms of physiological, anatomical, and social differences.

It was the German scholar Johann Friedrich Blumenbach (1752–1840) who gave the world its first racially based system of classification. After examining the skulls and comparing the physical characteristics of different peoples, Blumenbach concluded that humanity could be divided into five major races: Caucasians (West Asians, North Africans, and Europeans except the Finns and the Saami), Mongolians (other Asian peoples, the Finns and the Saami, and the Inuit of America), Ethiopians (the people of Africa except those of the north), Americans (all aboriginal New World peoples except the Inuit), and Malayans (peoples of the Pacific islands). These five divisions remained the basis of most racial classifications well into the twentieth century and continue to be commonly accepted in popular thinking even today. But population scientists now recognize the indefiniteness and arbitrariness of any such demarcations. Indeed, many individuals can be classified into more than one race or into none. All that can be said here is that the concept of *race* makes sense, if at all, only in terms of lineage: that is, people can be said to belong to the same race if they share the same pool of ancestors. But as it turns out, even this seemingly simple criterion is insufficient for rationalizing a truly objective classification of humans into discrete biological groups in such a way that everybody belongs to one and only one because, except for brothers and sisters, no individuals have precisely the same array of ancestors. This is why, rather than using genetic, anatomical, or physiological traits to study human variability, anthropologists today prefer to study groups in terms of geographic or social criteria. *Race* in the end is fundamentally a historical or cultural notion, not a biological one.

So, there really is no scientific basis to false notions such as the "master race" one, or other notions based on racial superiority or supremacy. But these persist nonetheless and can easily be manipulated by master liars to sow a sense of resentment and even hatred toward otherness. Discrimination is based on such manipulations, as can be seen in the Aryan myth. Origin myths and redemptive histories that are based on exclusion are dangerous to the very people who espouse them, as evidenced by the horrific failure of the Aryan myth. In attempting to bring a society back to a supposed period of racial purity, all they do is destroy that society, since there is no such thing as racial purity. In a parallel fashion to the Aryan myth, Mussolini attempted to bring back the glory of Roman times to his society by concocting a similar story of Roman purity, ignoring that many Italians did not have such heritage. In a statement he wrote to commemorate the founding of Rome, on April 21, 1922, he made the following assertion, linking ancient Rome to fascism:[35]

> Rome is our point of departure and of reference; it is our symbol, or if you like, it is our Myth. We dream of a Roman Italy, that is to say wise and strong, disciplined and imperial. Much of that which was the immortal spirit of Rome resurges in Fascism.

In fascism, not only the members of other races or ethnicities, but also "enemies from within" the superior race itself, are seen as dangerous to the establishment of cultural harmony and progress. In the case of the French and Bolshevik revolutions, the enemies were members of the aristocracy and their profligate lifestyle; in the case of fascism, the enemies were intellectuals and their antirealism approach to politics and society. The types of "inner enemies" may vary, but the strategy in any mythic origins story is the same—they are the ones who have helped bring about a destruction of society's real purpose and values by pandering to the politics of racial and cultural inclusivity. To restore society to its real historical mission, the politically correct and racially inclusive language and views of the liberal press and intellectuals must be attacked viciously through brutal slogans and clichés.

Slogans such as "enemies of the people" and the "deep state" resonate as a call to arms to eliminate the inner enemies, and may even lead to real arms-taking, as was evidenced by actual attempts in 2018 to attack Trump's "enemies." A case in point was that of a Trump supporter who had a "CNN Sucks" sticker on his van. He took it upon himself to attack Trump's enemies by sending pipe bombs to them.[36] These were perceived as the inner enemies, including liberal media personalities, past politicians who Trump had constantly denigrated, and others who supported the racial diversity worldview. Similarly, the body slamming of a *Guardian* reporter by a Republican congressman, and the plans for carrying out a mass murder if the enemies of the people by a member of the US Coast Guard who wanted to eliminate virtually anyone who opposed Trump,[37] are examples of how purity myths can fuel hateful feelings toward people who are felt to be enemies of the leader.

There is nothing more dangerous than stoking pent-up resentment against people of a different race or background. Trump's political rise was built on a convenient racist lie about Barack Obama's birthplace—the so-called birther conspiracy theory (chapter 1). This became a type of code, referring to the historical-cultural illegitimacy of an African American man as president of America. It would be naïve to think that Trump did not understand this subtext of the birther conspiracy. He was not simply lying—he was playing on the resentment that a descendant of the slaves had taken over the leadership of society. This type of unconscious hatred seems to bind many together. As Orwell so insightfully put it, "Let's all get together and have a good hate."[38]

The birther myth, as Orwell's quote suggests, brings out the reason why confabulation is so dangerous for the progress of a liberal democracy—it stokes hatred unconsciously. The myth is no longer stated overtly, having receded into the realm of the subconscious where it festers even more dangerously.

EPILOGUE

The rallies of Mussolini and Hitler were scary events. They told their racist myths over and over to adoring crowds. They knew how to make their falsehoods believable through the format of confabulation, utilizing highly emotional slogans and a body language that literally enthralled their audiences. The master liar-prince is also a master performer.

It comes as no surprise, therefore, to find that Trump too is a skilled orator and showman who knows how to deliver his orations effectively to adoring audiences through a crafty choice of the same slogans and discourse tactics of all kinds (some of which have been discussed previously). To ensure that his delivery will always have its desired effect, only those who are his fervent supporters are allowed into his rallies, guaranteeing the kind of revivalist atmosphere that reaches high emotional pitches at key points of delivery without any intrusive or mood-breaking opposition. During the primaries, Trump's audiences were mixed at first. As a result, some audience members became antagonistic to his stances, expressing their opposition vociferously, often approaching the stage to confront him. Trump counterattacked the opponents by denigrating them with slurs to the applause of his supporters. Soon after, only the latter were allowed into his audiences. This guaranteed that his performance would be maximally effective, stirring up resentments against the purported enemies of the real America.

As Machiavelli knew, appearances matter because people tend to look at the surface: "Men judge generally more by the eye than by the hand, because everybody can see you, but few can come close enough to touch you."[39] This is a central aspect in the enactment of the Art of the Lie, because, as Machiavelli goes on to say, "What you [the prince] appear to be, few really know what you are."[40]

The study of Trump's physical appearance and body language can tell us a lot about why he is so appealing to his supporters. When he is in front of his fanatical audiences at rallies, he assumes the same kind of body language of previous dictators, raising his head to the side as if looking into the heavens, recalling the same head tilt of Mussolini and Hitler, proclaiming his bluster imperiously and majestically. Trump's hair is also part of his performance persona. As semiotician Giampaolo Proni has perceptively put it, "There are few doubts that Trump's hairdo is surprising, unique, arousing curiosity and aesthetic controversial evaluations."[41] The orange-blond color evokes the symbolism of a "golden heroic warrior." As Arthur Asa Berger has also argued, Trump's hair has a mythological quality to it, comparable to Medusa's hair,[42] whose hair was turned into snakes as punishment for marrying Poseidon. A bald Trump would arguably have a diminished appeal. He always wears a suit and tie for his public appearances. Like a veritable patriarch, Trump aims to impart an aura of authority for himself and to carve out an image of a political leader who respects the formal business dress code of the "real America." His impeccable appearance, imperious poses at rallies, and his golden hair coalesce to make him appear like an ancient hero. His dress, appearance, and body language are all part of the act.

To conclude this chapter, it is useful to emphasize that confabulation is one of the more effective strategies in the Art of the Lie, because it stokes resentments and promises simple solutions to social problems. As B. Joey Basamanowicz and Katie Poorman have insightfully pointed out, by scapegoating immigrants and minorities Trump has produced a lethal, but deeply alluring, redemptive view of America that resonates with those who feel marginalized by liberals and intellectuals.[43] Confabulated histories tap into emotionally repressed beliefs and they are what make the Machiavellian liar so slippery to pin down, since everything he says and does is evaluated in terms of the perceived veracity of his redemptive story.

The attack on otherness is impervious to specificity—it can be African Americans, sexually diverse individuals, and immigrants (among others). Recognizing the rising sense of Islamophobia in America since 9/11, Trump targeted Muslims opportunistically from the outset of the campaign. In 2015, he stated bluntly during a speech that "Donald J. Trump is calling for a total and complete shutdown of Muslims entering the United States, until our country's representatives can figure out what the hell is going on, we have no choice. We have no choice. We have no choice."[44] Trump's statement that they needed to "figure out what the hell is going on" resonated with his audience since it was designed subconsciously to elicit images of 9/11 retrospectively and the hatred of America that these images evoked. The "complete shutdown of Muslims entering the United States" was thus put forth as the only way to stop the destruction, because "we have no choice." Of course, Trump has extended his vitriol to attack all those he labels as America's enemies (within and without), from Mexicans to the liberal media. Part of the overall strategy of identifying and weeding out the enemies is to create an enemies list—a tactic deployed by despots from Stalin and Mussolini to Richard Nixon. It thus should come as no surprise to find that Trump also established such a list, as Cliff Sims, a previous White House insider, wrote about in his book *Team of Vipers*.[45]

The question of why so many believe pseudo-histories and confabulations, such as the birther one, remains a true psychological conundrum. It seems, as journalist William Cummings of *USA Today* has aptly put it, "The human brain is wired to find conspiracy theories appealing."[46] Confabulations such as the Aryan myth involve the stimulation of mechanisms of belief in the brain, rather than those involved in rational understanding, as discussed in the previous chapter. As neuroscientist Antonio Damasio has cogently argued, the emotional areas of the brain often override its rational grasp of reality.[47] The limbic system—which includes portions of the temporal lobes, parts of the hypothalamus and thalamus, and other structures—has, in fact, been found to play a larger role than previously thought in the processing of certain kinds of speech and thoughts. In other terms, we might believe myths and conspiracy theories because they tap into our subcortical gut feelings, and no matter how many reasons are brought forth to contradict them logically, the limbic brain simply blocks them out. To cite Cummings again: "No matter how unlikely a given imagined conspiracy, and no matter how many facts are produced to disprove it, the true believers never budge."[48]

4
FAKE NEWS

Falsehood has an infinity of combinations, but truth has only one mode of being.

—Jean-Jacques Rousseau (1712–1778)

PROLOGUE

In one of his writings, author Norman Mailer stated that: "Each day a few more lies eat into the seed with which we are born, little institutional lies from the print of newspapers, the shock waves of television, and the sentimental cheats of the movie screen."[1] In these words, the roots of mistrust or, contrariwise, of overreliant trust in news organizations and the mass communications media can be discerned. Realizing the power of the press to evoke trust or mistrust, liar-princes like Mussolini, Hitler, and Stalin attacked the press constantly, brow-beating liberal journalism into compliance with their caustic attacks. Hitler controlled the press and all broadcast media by establishing the Reich Ministry of Public Enlightenment and Propaganda, which enforced Nazi ideology on journalists and media personalities. It was Hitler's version of Orwell's Ministry of Truth. In Soviet Russia, censorship of the press and the electronic media was institutionalized throughout the Soviet Union. *Pravda*, a daily newspaper, was founded in 1912 to ensure that the press played along with the party line. From 1918 to 1991, it was the official organ of the Soviet Communist Party.

Trump's constant attacks on CNN, the *Washington Post*, *The New York Times*, and other liberal media as "enemies of the people" who spout "fake news" falls into the same category of attack on the free press witnessed in the regimes of Mussolini, Stalin, and Hitler. On the other hand, Fox News, tabloid, and alt-right social media are analogous to the state-controlled media of the Mussolini era. Trump's clever strategy of calling the media that critique him "fake news" and those supporting him "real news," including sensationalistic internet sites such as *Infowars*, does not emerge in a vacuum. It is actually consistent with a longstanding journalistic tradition in America that can be traced back to the advent of so-called yellow journalism in the nineteenth century. The sensationalistic media outlets of today are really the descendants of yellow journalism. The strategy of calling news outlets such as CNN and MSNBC "fake news" is a salient attempt to undermine critical coverage of Trump. It is not surprising that, like previous autocrats, he has constantly called for government control and even censorship of these outlets. Perhaps he envisions the Federal Communications Commission as his personal Ministry of Truth.

As in other areas of linguistic manipulation, Trump knows that his attack on the media will be taken at face value by his base, since such media are portrayed by him to be a propaganda arm of the deep state—a nefarious reverse psychology strategy of blaming the attackers for what he does himself. With the "fake news" slogan, he is able to kill two birds with one stone—to attack his attackers and to promote the conspiracy theory that he is being victimized by the "enemy." From this situation, a media-based "fight for the truth" has become an ongoing one that pits styles of journalism against each other—tabloid versus serious—in an attempt to gain the upper hand. Since Trump's election, a "fake news syndrome," as it may be called, has emerged and become widespread, given the migration of news reporting to the infinite universe of social media where conspiracies fester and spread, leading to a "post-truth" era of journalism and a weakened democracy as a result. The fake news syndrome can be defined as the perception that one's preferred news source is truthful, while others, who present the news in ways that are in contrast with it, are seen as untruthful and self-serving. No wonder, then, that news outlets like CNN have made the slogan "Facts matter" part of their response to Trumpian doublespeak.

It is quite extraordinary, actually, to contemplate that yellow journalism is ultimately the source of the current fake news syndrome that is afflicting politics and society. What those yellow journalists knew, and what master liars and despots have always understood, is that people believe information that is presented to them in sensationalistic ways, especially if it is consistent with conspiratorial beliefs (as discussed in the previous chapter).

The goal of this chapter is to examine the "fake news syndrome" and its effects on human minds. Indeed, it is no exaggeration to say that it may be negatively affecting the mental health of many. The constant twisting of facts for self-serving purposes and the assault on truth have emotionally destabilizing effects. Since today we get most of our news from a plethora of social media, we are being constantly bombarded with fake reports and conspiracy theories. It is relevant to note that the origins and spread of fake news during the 2016 presidential election have been traced to websites in Veles, a small town in Macedonia, where a group of teenagers made substantial money through Facebook by writing completely fictitious stories about American politics, most of which were intended to dupe people into seeing Donald Trump as the only viable president.[2] The cluster of sites operating out of Veles brought serious attention to the role of social media in bringing about a post-truth world, in which disinformation can come from anyone, anywhere, rather than just Ministries of Truth, as was the case in the not-too-distant past. The spread of falsehoods is now just an algorithm away. In cyberspace, it is virtually impossible to separate the fake from the real, facts from alternative facts, real stories from confabulations.

ORIGINS

"Fake news" is defined as a deliberate form of disinformation spread via news media outlets on different platforms. The term *deliberate* is crucial here, since mistakes in the interpretation and presentation of facts occur all the time, but they are not deliberate or premeditated. When such faux pas or gaffes occur on legitimate news media, they are acknowledged as such and relevant corrections are announced. It is on yellow journalistic media (past and present) that the disinformation is planned and ingeniously contrived. When exposed or even just challenged, the disinformation is never corrected by such media, but rather adjusted to justify and double down on the falsehoods. As communications analyst Piero Polidoro has aptly observed, the rise and spread of the fake news syndrome did not occur in a cognitive void. It traces its roots to political, social, technological, and cultural forces that converged in the nineteenth century, sparking a society-wide need to find quick answers to complex problems, as well as a distrust of the views of traditional authoritative institutions.[3] This need and distrust have spread throughout cyberspace, where the tendency to accept information at face value, without the conscious deployment of critical interpretive filters, is a widespread habit of mind. Even those who do apply the filters are likely to be negatively influenced by the massive proliferation of fake news. We all have a saturation point in the assessment of information after which we start to ignore the implications of the information at hand.

Social historians pinpoint the emergence of tabloid-style journalism to Randolph Hearst (1863–1951), an influential figure in the history of American journalism. Hearst employed circulation-boosting tactics, such as lurid and sensationalized headlines and stories, along with unverifiable gossip articles about celebrities and well-known media and political personages. The term *yellow journalism* was coined in the 1890s in reference to the rivalry between two New York City papers, the *New York World* and the *New York Journal*. The techniques used by both included banner headlines, flashy illustrations, and a section of comic strips. The name actually comes from the comic strip developed by American cartoonist Richard Felton Outcault (1863–1928), called *Hogan's Alley*, which first appeared on May 5, 1895, in the *New York Sunday World*. The strip depicted the underside of the city with decrepit tenements, backyards with dogs and cats running around randomly, tough guys, urchins, and ragamuffins, who were always involved in mischief or wrongdoing. One of the urchins was a flap-eared, bald-headed child with a quizzical yet shrewd smile named Mickey Dugan. He wore a long yellow gown. As a result, he came to be known as the "Yellow Kid." In 1896, *Hogan's Alley* was in fact renamed *The Yellow Kid*. The popularity of the strip initiated a newspaper war, with all kinds of newspapers attempting to outdo the popularity of the strip with their own cartoon characters and with hyperbolic language. The term *yellow journalism*, meaning a sensationalistic style of newspaper writing, was derived from Outcault's comic strip.

Yellow journalism migrated to the early tabloid newspapers and magazines, which featured stories on the occult, crime, disaster, scandal, gossip, media celebrities, and unfounded conspiracy theories, from political to scientific ones. It spawned the fake news syndrome, introducing a simplified language that was immediate and visceral, designed to grab attention. Already in 1835, long before social media conspiracies, an article in the *New York Sun* reported the presence of life on the moon, which was, needless to say, fake information intended to play on people's love of mystery and their fascination with unexplained phenomena, fueled by the common belief that the government was hiding the truth about such things as extraterrestrial life. The article claimed to show plants and other forms of organic life on the moon's surface. Many were dumbstruck by this misinformation. The article was discussed in other news media, leading to debates on extraterrestrial life without any science to support any of the claims.[4] Remarkably, even in this preinternet era, the fake information spread like wildfire not only to other news outlets but also at workplaces and schools. The story had become larger and larger, with new elements being added cumulatively to its plot line, evolving into what we would today call a meme or an urban legend. It is not clear if the fake information was a parody of science or a prank perpetrated by the editor, Richard Adams Locke, on a gullible readership, eager to read about conspiracies that tapped into a reflex distrust of the government and a thirst for stories of alien life. The moral of that story (pun intended) is that we are all susceptible to being deceived if the story is interesting, if the language is persuasive and immediate (much like a mystery novel), and if it tells us something we suspect may be hidden from us.

The current fake news syndrome has a long history behind it. Or, to use a medical metaphor, it has an identifiable etiology, defined as the causation of a disease or disorder. The metaphor is an apt one for characterizing how misinformation affects the mind negatively. Fake news and conspiracy theories are capable of directly warping minds with the same kind of impact a disease or epidemic has on the body. The symptoms of the fake news syndrome can be devastating, not only on individuals but also on entire societies, since falsehoods spread throughout a community like a virus, as the fake moon story showed.

Today, fake news is delivered mainly through social media and internet disinformation campaigns, such as the Russian hacking of the 2016 presidential elections, whereby Russian hackers spread falsehoods through Facebook, playing on racial divisions and resentments in America. The believability of the disinformation was bolstered by images of race riots and social disorder that were jarring and highly suggestive that drastic changes were required—pointing to Donald Trump as the "effector" of the changes. The fake news was so realistic that many of those who were influenced by it, when interviewed subsequently, claimed that the stories they read could not possibly be fake. Trump supporters simply interpreted the disinformation as real, or at least plausible, since the images evoked fears subtly that riots and disorder would continue unless radical political change was brought about. What this illustrates is that once a conspiracy theory is promulgated, tapping into suppressed beliefs, it spreads its conceptual roots deeply into the mind, filtering out all subsequent contradictory information.

The fake news syndrome has spread even more broadly during Trump's presidency. An interview of a Trump supporter after the infamous 2018

Helsinki conference, where Trump stated his support for Putin over his own intelligence agencies, is a case in point of how the syndrome distorts the perception of reality.[5] The supporter acknowledged that the Russians had indeed interfered with the elections, adding the following: "It's nothing new. They've been doing it for years. And [Trump] didn't look at Putin and say, 'Hey, you're lying.' He negotiates different than every other politician." As the interview progressed, it became obvious that the subject perceived all that Trump did as part of a larger strategy to change America for the better.

There is, in actual fact, an ever-expanding and lucrative fake news industry today that has found a fertile ground in cyberspace. Fake news stories that go viral on social media can draw significant advertising revenue when users click on the original website. Some websites specialize overtly in fake news, knowing that those who follow them are impervious to the falsehoods, since they are perceived as advancing the political agendas of their favorite candidates and as validating their own conspiratorial beliefs. As the fake news spread throughout the internet, they are shaped and reshaped over and over by anyone online, thus creating a veritable vicious cycle among the consumers and creators of fake news. This constant exposure to falsity has a kind of Pavlovian conditioning effect that gradually erodes the ability to objectively assess the accuracy and quality of information. If the information is consistent with beliefs and ideological agendas, then, whether true or not, it is virtually impossible to get the consumer to reject it as mendacious.

This syndrome is particularly marked in those who were born, and have been raised, in the era of the internet. In a 2015 study of thousands of students ranging from middle school to college, researchers at the Stanford Graduate School of Education concluded that the online generation can be easily duped with disinformation if it occurs on social media sites such as Facebook and Instagram.[6] Many of the respondents were, seemingly, unable to distinguish fact-based news from their fake counterparts, regardless of their technical savvy. If this study is repeatable even in minimal ways, then it would lend support to the validity of the theory that there are consequential psychological repercussions wreaked by the fake news syndrome. Falsehoods can easily spread through the digital universe with minimum effort and low investment of time and resources on the part of the conspirator or schemer, who is engaging, essentially, in a facile form of mind control. Free speech laws prevent the prosecution of the creators of fake content. Disinformation, clickbait, hoaxes, conspiracy theories, pseudoscience, and bogus content are so dominant today that they induce what can be called, figuratively, a catatonic processing of information, rendering critical thinking virtually immobile or unresponsive. We live in the perfect cognitive environment for the liar-prince to thrive in.

As discussed in the previous chapter, once a mindset is reshaped by deceptions and lies, only time and circumstances can reverse the impact on the mind that the liar has perpetrated. Mussolini, Stalin, Hitler, and others were not brought down by cogent counterarguments, but by circumstances (economic, military, and social) that they could not avert with mendacity alone. Only when people realize that the "emperor has no clothes" through happenstance will the liar-prince be defeated.

In a classic treatment of the media, titled *Manufacturing Consent*, social critics Noam Chomsky and Edward Herman put forth a theory of how news media can guide and even manufacture consent on all kinds of issues, from politics to morality.[7] The ways in which the media present and package news coverage caters to the brokers who control the funding and (in many cases) ownership of the media. As a consequence, the media tend to be a vicarious propaganda system, set up to manufacture consent that is consistent with that of the brokers. They do this by selecting the topics to be showcased, establishing the tone of the issues that are discussed, filtering out contradictory information, or else challenging it in some argumentative way.

Now, because of the internet, virtually anyone can propagate fake news in the service of a political agenda. Some radical ultraconservative digital news outlets, for instance, aim to instill a view of the world that espouses an elemental form of patriotism and the essential benevolence of those who support their view. Herman and Chomsky describe five filters in the consent-making process. The first one is the ownership filter, implying that news information is filtered (behind the scenes) by those in charge to control the content and to present it accordingly. The second is the funding filter, whereby media outlets generally tend to subscribe to the views of those who fund them. Today, the funding can come directly from audiences themselves, who support the cause espoused by a news outlet through contributions. The third is the sourcing filter, by which interested parties are the ones who actually provide the news to be broadcast to the media outlets, filtering out, and even censoring, those not deemed to be supportive of their own views. The fourth filter is flak, which Herman and Chomsky define as any negative reaction to news items—a result that is to be avoided as much as possible. The fifth filter is the "anticommunism" one, which is the view that any political ideology contrary to the dominant one must be repressed or portrayed in negative terms.

OCCULTISM

One of the more popular features of yellow journalism was its incorporation of occultism, with its various columns and rubrics on astrology, horoscopes, and mythic beliefs. To this day, the horoscope is a major feature of tabloids and other kinds of newspapers.

An example of an online site that perpetrates fake news using occultism as a lingua franca is *Infowars*—rather appropriately titled—which is an alt-right website founded in 1999 by rabble-rouser journalist Alex Jones, operating under Free Speech Systems LLC. Because of its promotion of hate speech, the site was suspended from various internet venues, including YouTube and Facebook in 2018. Jones has also been accused of sexual harassment by employees of the site. So, as I write, *Infowars* has become weakened as a fake news perpetrator. But during the 2016 presidential election, it was a major force in manufacturing consent for the Trump presidency. The site was especially famous for its conspiracy theories, including the unfounded claim that the 9/11 attacks were presented falsely by the mainstream media. Jones has had to retract some of his more vile conspiracies as a result of legal challenges. The question of why so many clicked on and accepted *Infowars*'s version of the news, during the presidential campaign, provides a key insight into how yellow journalism can shape beliefs. Jones constantly generated a state of fear of something lurking behind the scenes—a shadow state—that was out to get Trump. The term was derived from occultist language, not in the mystical sense of the word, but in the sense of something that is beyond the realm of ordinary knowledge and to which Jones had exclusive access.

It is all too easy to dismiss someone like Jones as a quack. But his style and approach are grounded in the tradition of yellow journalism. As Richard Wooley has aptly observed, Jones and Trump are contemporary figures in America's fascination with sensationalism, conspiracies, occultism, and outright historical falsehoods—a compulsion which people have been able to indulge since the advent of yellow journalism and other forms of sensationalistic reportage:[8]

> Yellow journalism is part real news, part confabulation, and a large part entertainment. It translates the *esprit* of the circus into print. Because of globalism, this is not just an American problem, especially in cyberspace, which knows no political or national boundaries and where sensationalism has crystallized as the Esperanto of communications. The type of journalism on sites such as *Infowars* is part of a circus-type spectacle, an example of yellow journalism-as-entertainment.

Since the outset, yellow journalism understood the emotional power of occultism, or the belief that mythic, mysterious forces rule the universe—

hence its adoption of arcane occultist themes and rubrics (such as horoscopes), which are used in place of science to explain all kinds of physical, social, and psychological phenomena. This is why yellow journalist pundits like Jones reject science outright, critiquing it as a self-serving conspiracy of its own that rejects any and all contrary views to its theories, such as his own occultist ones. Occultism is part of *mythos* and can thus be traced back to the origins of human cultures. Ironically, it was an obscure but important presence in the religious Middle Ages. Even eminent Church figures, such as thirteenth-century Italian theologian Saint Thomas Aquinas, believed in the powers of alchemy and other occult arts. The late medieval and early modern period saw occultism increasingly as being connected with worship of the devil. For this reason it was censured, resulting in the persecution of "witches" during the Renaissance. However, by so doing, the Church really breathed new and exciting life into occultism, which spread in Europe in the eighteenth and nineteenth centuries, adopted in part by the artists and composers of the era, who saw great value in occult and mystical traditions because they revealed the creative powers of the imagination. Occultism surfaced again in the New Age and counterculture movements that gained momentum in the 1960s and 1970s in America. Trump's slogan of the "witch hunt" reverberates with occultist connotations, at the same time that it is a means to attack his opponents.

As Gary Lachman has argued in his perceptive book, *Turn Off Your Mind*, occult beliefs today have become common because they were adopted not only by yellow journalists but also by the counterculture youths of the 1960s, 1970s, and 1980s.[9] Bands and artists introduced (or more correctly reintroduced) everything from the tarot, astrology, yogis, witchcraft, and other forms of occultism into America, where they remain firmly entrenched to this day. And this has had a profound effect, he claims. The movie *The Matrix*, for instance, has led (according to Lachman) to the rise in brutal serial murders, suggesting a possible osmotic effect between occult symbolism and real-life horror. Lachman puts it as follows:[10]

> The rise of seemingly pointless serial killings gives pause for concern. Likewise the horrific happenings at the Columbine High School near Denver, Colorado, when Eric Harris and Dylan Klebold shot dead a dozen of their fellow students. Dressed in black raincoats, the two casually slaughtered their classmates, before turning their guns on themselves. It later turned out that they had devised a plan for even greater destruction, including hijacking a plane and crashing it into a major city. If the killings weren't macabre enough—and although there's no causal link, both were fans of various "shock rockers"—they seemed eerily paralleled in a hit sci-fi film of the time, *The Matrix* (1999), in which Keanu Reeves, guns-ablaze, leads a band of black leather-clad psychic hackers out of the prison of a false reality. The Gnostic motif of breaking through to the other side had a mini-renaissance in some late-nineties sci-fi thrillers, like *Dark City* and *The Cube*. But in *The Matrix* this theme is coupled with a Gestapo-like dress code, shades and plenty of guns. Dark glasses, leather coats and automatic weapons met the ancient Gnostic dream of escaping the prison house of the flesh. Magic is still alive today. It is just that its practitioners don't all wear sandals.

The need to "escape the prison house of the flesh," as Lachman puts it, has often been satisfied by occult practices and traditions. It should come as little surprise, therefore, that websites that perpetrate fake news, such as *Infowars*, are deeply engaged in occult conspiracies and why these resonate unconsciously with so many people. As emphasized throughout this book, belief can be made to overwhelm reason through language and narrative content. It is one of the judicious lessons to be learned from history.

COUNTERATTACKS

Trump's constant counterattacks on what he identifies as the fake news media—which are those that expose his falsehoods—creates a confusion of mind in which the facts can be twisted over and over for self-serving purposes, allowing him to portray himself as a victim rather than a victimizer. Trump knows how to strategize the occultist fake news atmosphere in which we live to his advantage. The fact that some conservative media outlets, such as Fox News, adopt his false victimization pronouncements and shape them into believable stories is truly mind-boggling, since they are often demonstrable as being patently false or self-contradictory.

The following tweets are examples of how Trump shrewdly manufactures his victimization. They were written following his first meeting with the North Korean dictator, Kim Jong-un, in 2018. In them he portrays himself as a victim of news coverage, creating confusion by intersplicing his victimhood into his claims of achievement:

> If President Obama (who got nowhere with North Korea and would have had to go to war with many millions of people being killed) had gotten along with North Korea and made the initial steps toward a deal that I have, the Fake News would have named him a national hero![11]

> Many good conversations with North Korea—it is going well! In the meantime, no Rocket Launches or Nuclear Testing in 8 months. All of Asia is thrilled. Only the Opposition Party, which includes the Fake News, is complaining. If not for me, we would now be at War with North Korea![12]

This is not only a self-aggrandizing strategy—typical of Machiavellian liars—but also a way of counterattacking the news media that are critical of him. Another salient strategy inheres in the propounding of false, bald-faced accusations against the fake news media, for which Trump provides no evidence to back them up, as can be seen in the two tweets below in which he attacks the media for giving Obama more airtime than they give him, portraying himself, again, as a victim of the fake news media:

> Such a difference in the media coverage of the same immigration policies between the Obama Administration and ours. Actually, we have done a far better job in that our facilities are cleaner and better run than were the facilities under Obama. Fake News is working overtime![13]

His accusations of the mainstream media for not supporting his skewed and self-serving immigration policies can, by themselves, constitute a textbook treatment of how to use the victimization strategy to undercut the reportage of opponents in effective Machiavellian style:

> The Fake News is not mentioning the safety and security of our Country when talking about illegal immigration. Our immigration laws are the weakest and worst anywhere in the world, and the Dems will do anything not to change them & to obstruct—want open borders which means crime![14]

Perhaps Trump's most effective strategy is to boast about himself, emphasizing how the fake news media downplay his achievements:

> The Fake News Media is desperate to distract from the economy and record setting economic numbers and so they keep talking about the phony Russian Witch Hunt.[15]

> So funny to watch the Fake News, especially NBC and CNN. They are fighting hard to downplay the deal with North Korea. 500 days ago they would have "begged" for this deal—looked like war would break out. Our Country's biggest enemy is the Fake News so easily promulgated by fools![16]

Trump's constant questioning of the justice system before he appointed William Barr as his Sheriff of Nottingham attorney general is yet another self-serving counterattack strategy, intended to disseminate the false belief that there is a swamp in that system that is after him, populated by liberals, the mainstream media, Democrats, and anyone else who opposes him. Along with CNN, NBC, ABC, and CBS, he shrewdly portrays justice agencies such as the FBI as part of that swamp:

> Reports are there was indeed at least one FBI representative implanted, for political purposes, into my campaign for president. It took place very early on, and long before the phony Russia Hoax became a "hot" Fake News story. If true—all time biggest political scandal![17]

> NBC NEWS is wrong again! They cite "sources" which are constantly wrong. Problem is, like so many others, the sources probably don't exist, they are fabricated, fiction! NBC, my former home with the Apprentice, is now as bad as Fake News CNN. Sad![18]

> The Fake News Networks, those that knowingly have a sick and biased AGENDA, are worried about the competition and quality of Sinclair Broadcast. The "Fakers" at CNN, NBC, ABC & CBS have done so much dishonest reporting that they should only be allowed to get awards for fiction![19]

As is saliently evident, Trump's tweets employ the same sensationalistic style of yellow journalism that is meant to inculcate his views effectively. The elements of this style include the use of capital letters, exclamation marks, succinct remarks, slang, terse slogans, and hyperbole. Reading Trump's tweets is akin to reading circus posters or tabloid headlines, which are designed to attract people through lurid and garish jargon. It constitutes a speech style that is not inconsistent with the Protestant ethic discussed previously—a form of harmless escapism from the troubles of the world through garishness. As the American poet e. e. cummings aptly put it: "The tabloid newspaper actually means to the typical American what the Bible is popularly supposed to have meant to the typical Pilgrim Father: a very present help in times of trouble, plus a means of keeping out of trouble via harmless, since vicarious, indulgence in the pomps and vanities of this wicked world."[20] Trump's forceful counterattack tweets "indulge" his adoring audiences, while confusing and frustrating his critics at the same time.

THE FAKE NEWS SYNDROME

Why are we all so prone to believing falsehoods, at least sometimes? Is the answer to be found in the history of yellow and tabloid journalism as discussed above, bolstered by the online media that produce false beliefs by virtue of the fact that whatever they showcase, and how they do so, becomes part of general credulity created by what has been called here the fake news syndrome? The answer to such questions is a qualified yes. This was brought out by the Cantril study, discussed briefly in the opening chapter.[21] To recall it here, the study aimed to understand why the 1938 radio broadcast of a CBS play based on H. G. Wells's *War of the Worlds* created such panic among many listeners. The broadcast interspersed fake news reports of Martian landings in New Jersey, which were so realistic that many listeners believed that they were real, despite periodic announcements that the program was merely a fictional dramatization. As indicated previously, the Cantril study found that better-educated listeners were more likely to recognize the broadcast as fake than less-educated ones. But the study may have missed the unconscious source of the hysterical reaction—namely, the fake news syndrome that spreads like an emotional virus in a segment of the population that has become so accustomed to sensationalistic fake media reportage that it can no longer differentiate between reality and fiction.

The fake news syndrome is now spreading like wildfire via cyberspace, where falsehoods are concocted routinely. This syndrome can thus be connected to a larger simulation syndrome—a psychological simulacrum, as mentioned previously, whereby fictional representations mesh with reality to become indistinguishable from it. This is, actually, perceived by many as a desirable feature of social media, allowing anyone to portray oneself through the simulacrum. The first social network site, SixDegrees (sixdegrees.com), which was launched in 1997, introduced this kind of simulation broadly by allowing users to create self-styled profiles and "Friend Lists." It closed down in 2000, perhaps because the idea of a social media community was too new at the time. Others emerged shortly thereafter, expanding the requirements of membership, such as allowing combinations of profiles, viewable materials, guest books, and diary pages. In 2000, LunarStorm, a Swedish networking site, added guestbooks and diary pages, which have become common features of today's social networking sites. From 2001 to 2002, several important social networking sites were launched, including LinkedIn and Friendster. Friendster was transformed, initially, into a dating site, designed to help "friends of friends" meet each other through viewable profiles. As a consequence, Friendster surged in growth. In fact, because Friendster's database and servers were unable to handle the exponential growth, the site would often crash. Combined with frequent "Fakesters," people who outright pretend to be someone they are not, such as a celebrity, and the company's response of deleting all users with fake photos, many lost trust in the site and abandoned it en masse. It became saliently obvious with the advent of Facebook and later of YouTube, Instagram, and other platforms that the simulacrum had become a kind of unconscious strategy of online construction of the self.

Today, social media platforms such as *Infowars* are fake news factories, as media analyst Mark Dice has cogently argued.[22] They spread false information as part of political agendas through conspiracies and other kinds of falsehoods. As Dice claims, the fake news stories that went viral during the 2016 campaign, perpetrated by Russian disinformation hackers, undoubtedly affected the outcome because of the confusion that the disinformation was designed to create. The false messages were contrived in such a way that they evoked a sense of malaise in society attributed to the politics of diversity and the state of affairs that it had generated. Facebook was the main site for spreading the disinformation, simply because it was the most extensively used social network in America at the time. Like village gossip, the contents of Facebook pages are felt to hide truth. A perusal of postings on Facebook sites shows that people do indeed use the medium to gossip about others or to defend themselves against gossip. Clearly, social networking systems have redefined the nature of interaction, communications, and social relations.

The socio-philosophical implications of social networking have been studied from many disciplinary angles, since they are immense. These need not concern us here. The relevant point is that the degree to which the fake news syndrome is now spreading is a consequence of the social media universe in which we live. It is part of virality and the memetic structure of internet content. This phenomenon cannot be easily explained or categorized by traditional theories of media and culture. An anonymous musician playing classical music in a clip may be viewed over sixty million times; an inebriated person might also get millions of hits; a cat playing the piano goes viral throughout the world; and so on. It is no coincidence that most online fake news reportages tend to be short pieces based on humor or surprise. A long-winded piece of news reportage would not attract anyone who would visit such sites. The fake news syndrome in online memetic culture is a key to understanding the nature of the political "populism" that has taken hold today—a populism based on the language and style of yellow journalism that has been adopted and spread even more broadly through cyberspace.

DISINFORMATION

Disinformation can be defined for the present purposes as a strategic spread of fake information for political or ideological purposes, intended to deceive people or else reinforce or distort their existing beliefs (as the case may be). The first modern-day use of disinformation tactics can be traced to Soviet Russia under Joseph Stalin, who coined the term itself and founded a "Disinformation Office" in 1923—his version of the Orwellian Ministry of Truth.[23] In the post-Soviet era, the disinformation strategy has hardly evanesced, since it is continued as a key military and social engineering tactic under Vladimir Putin, who used it effectively during the 2016 American presidential campaign, spreading falsehoods through social media platforms with the use of bots, or autonomous programs that can interact with computer systems or users. The success of Russia's disinformation tactics can be seen, arguably, in the mind-controlling effects it has had on American politics and other nations. The intent is destabilization through disinformation. The main tactic is to construct positive information to coddle and dupe social media users themselves so that the deceptive information will fall below the threshold of awareness.

It is evident that many of Trump's tweets fall under the disinformation rubric. They are so blatant and nakedly self-serving that linguists and social scientists are trying to understand why they are so effective (including the present one). His mode of disinformation has itself become a new type of Machiavellian discourse, adopted and promulgated by the alt-right fanatics of news outlets such as *Breitbart News*, and even, to some extent, Fox News. As a result, millions of Americans now believe that *The New York Times*, the *Washington Post*, CNN, and MSNBC, among other Trumpian fake news targets, are seditious institutions (enemies of the people) bent on removing from office the person they elected.

Actually, this strategy has backfired or has, at the very least, been counterproductive. Never before in the history of the press have *The New York Times* and the *Washington Post* emerged as the bold and unswerving voices of justice and truth and their journalists as the brave fighters against disinformation, attacking abuses of political power, recalling their similar role during the time of the Pentagon Papers and Watergate.

As discussed throughout this book, there are various historical parallels to the present situation. One was the rise of Benito Mussolini to power that was aided by his strategic use of disinformation. At thirty-nine, he was the youngest ever Italian prime minister. His rise was undoubtedly bolstered by his charismatic style of oratory that mesmerized his audiences with his supercharged slogans. He used disinformation and rhetorical bombast to build an undemocratic, authoritarian, and repressive state, taking advantage of a country that, at the time, lacked unity, proposing nationalism as a unifying force. Mussolini was perceived *as* Italy. He rejected the government at the time as "a gathering of old fossils."[24] Gaining the trust of the people, he then proceeded to strip away the rights of the free press. In July 1924, Mussolini justified his takeover of the polity by claiming that it was the people who had asked him to do so, so that he could restore social order:[25]

> The people on the innumerable occasions when I have spoken with them close at hand have never asked me to free them from a tyranny which they do not feel because it does not exist. They have asked me for railways, houses, drains, bridges, water, light and roads.

On January 3, 1925, Mussolini made the following proclamation: "I and I alone assume the political, moral and historic responsibility for everything that has happened. Italy wants peace and quiet, work and calm. I will give these things with love if possible and with force if necessary."[26] It is unnerving to witness that in a North Carolina speech, Trump made the following similar type of statement, referring to the purported mess that the world he "inherited" was in: "I will fix it. Watch. I will fix it."[27] As an editor for several newspapers himself, Mussolini had learned the art of disinformation firsthand, coming to use the press for his own ends. Shortly after his takeover of power, most of Italy's mainstream newspapers were suppressed, with a few smaller ones claiming to be independent tolerated in order to give the appearance of freedom of the press, but they were a smokescreen designed to cover the obliteration of such freedom. Without any challenge, Mussolini's megalomania flourished, as the crowds at his rallies cried "Duce, Duce, Duce! We are yours to the end."

A similar disinformation approach to politics came, needless to say, with the rise of Nazism in Germany in 1933, when Joseph Goebbels was appointed as the minister of propaganda on March 14 of that year. The first thing that Goebbels did was to spread disinformation on Germany's descent into social and economic chaos, claiming that it was engineered by an attack against white, educated young people in Germany. This allowed Goebbels to recruit party zealots who were young and smart and who displayed "ardor, enthusiasm, untarnished idealism."[28] Goebbels, like Mussolini, took control of the press, labeling Germany's newspapers as "messengers of decay" that were injurious to the "beliefs, customs and national pride of good Germans."[29] As a result, all journalism was subjected to *Gleichschaltung* (the standardization of political, economic, and social institutions in terms of Nazism)—implying that all journalists had to follow Nazi ideology on all issues.

The parallels between Fascist Italy, Nazi Germany, Soviet Russia, and current attempts at discrediting the mainstream press in America by radical conservative elements are as striking as they are scary. The difference between the latter and all the former is the fact that there is no need for Disinformation Offices or Ministries of Truth today—social media will do the same job rather well, having become part of a cultural war, with political operatives publishing disinformation to smear opponents and institutions, creating confusion and general discombobulation, which was one of the central aims of all Ministries of Disinformation of the past.

CONSPIRACY THEORIES

During a 2017 interview, Trump's advisor Kellyanne Conway attributed the "Bowling Green massacre" in Kentucky to the terrorist actions of two Iraqi men, whom she claimed were radicalized by ISIS and who came to the United States on purpose to carry out the massacre.[30] But no such event had occurred. When she was confronted with this fact, she characterized her fake story as a simple "mistake." It was hardly just a simple error; it was an attempt to spread disinformation surreptitiously as part of a larger conspiracy theory that paints Muslims as dangerous. Two Iraqi men were in fact arrested near Bowling Green, but years prior to Conway's conspiracy theory and for different reasons. The attempt to obfuscate actual facts with distortions is the main characteristic of conspiracy theories.

Conspiracy theories fabricated to support political causes perversely have occurred throughout history. By and large, they are based on the premise that nothing happens by accident and that what we see is not what is really true—thus rendering them unfalsifiable by rational counterarguments. Machiavelli saw the concoction of such theories as a critical tactic in gaining and maintaining political power, writing an entire treatise on conspiracies to emphasize their important role in the repertoire of mendacity that the liar-prince must exploit.[31] As Alessandro Campi notes, Machiavelli had developed a practical manual for how to plot a successful coup d'état or seizure of power.[32] The objective of this tactic is to generate paranoia among as many people as possible, which then spreads like a virus to infect virtually everyone, even those who do not believe the conspiracy but may find themselves needing to dispute its mendacity, thus adding to the cumulus of paranoia that the conspiracy is designed to evoke. The subtle art of political conspiracy involves, in other words, creating a false narrative in which enemies of the state are portrayed as scheming against it, either from outside or within it, manipulating reality in the process. From this paranoid swamp the Machiavellian prince emerges as a political savior that has come to the rescue (to drain the swamp). The philosopher Karl Popper argued cogently that totalitarianism is founded on conspiracy theories since they aim to pit people against each other in a subtle way that induces fear-mongering and paranoia.[33]

Since the nineteenth century, the main medium for divulging and diffusing conspiracies in America has been the yellow press and its descendants—the tabloids and online media (from radio to the internet). *Infowars* was, again, a perfect example of a conspiracy-based talk show whose sole purpose was to create confusion and sow paranoia in its listeners, attacking the liberal press and politicians as the enemies from within. *Infowars* was nothing new. The type of talk show it exhibited has always been a source of conspiracies, going back to 1926, in the early days of radio, when Father Charles E. Coughlin hosted a weekly talk show that attracted forty-five million listeners that dealt in conspiracies from program to program.[34] The underlying objective of the show was to promote conservative religious causes via the conspiracies. It was one of the first highly successful radio programs in the history of American media.

Talk shows now abound, with hosts, such as conservative media pundit Rush Limbaugh, promoting conspiracies on a daily basis. Of course, progressive or liberal talk shows, devoted to liberal politics, have also been around for a considerable period of time. These emerged in the 1960s, on progressive radio stations such as WMCA in New York and WERE in Cleveland, featuring outspoken hosts such as Alan Berg and Alex Bennett, who espoused liberal views of controversial events such as the Vietnam War and civil rights. The first cycle of TV talk shows started in 1948, and it is still part of the television landscape today, with talk shows interspersed throughout the day and night. These reflect the political diversity of America, with their differing political stances, from liberal to conservative and alt-right. Podcasts have emerged to bolster the popularity of the talk show format.

There are three plausible reasons why online talk shows have become particularly effective in spreading conspiracy theories and fake news. First, the cost of entering the market and producing material is little or nonexistent. This increases the pool of those who would promote fake news. Second, the nature of social media networks themselves makes it highly difficult to adequately judge the legitimacy of a news item—whatever is online can be construed in terms of varying degrees of falsehood or truthfulness depending on how the news is packaged and who the reporter or talk show host is. Third, high levels of ideological segregation exist among social media users. This makes them more prone to read and share similar politically aligned items that "stick" in minds, as Chip and Dan Heath have cogently argued, who identify six qualities that make ideas "sticky"—simplicity, unexpectedness, concreteness, credibility, emotions, and narrative coherence.[35] These are self-explanatory and need little commentary here. To this, I would add repetition—that is, by repeating the same conspiracies and falsehoods over and over, the fake news syndrome kicks in, affecting people's minds enduringly.

Because of this syndrome, Trump is in an unassailable position, with the liberal media playing right into his hands by attempting to debunk his falsehoods. Followers and allies believe him, or at least see his lies as necessary. What happens most of the time is a chase after the truth that is exasperating because it leads nowhere. This creates a kind of "catch-22" predicament, a dilemma from which there is no escape because of mutually conflicting stories that cannot be verified or falsified. As is well known, the origin of that expression comes from the title of a 1955 novel by Joseph Heller,[36] in which the main character feigns madness in order to avoid dangerous combat missions, but his desire to avoid them is taken as actual proof of his sanity.

As Geoff Nunberg has observed, the deep state conspiracy reflects how fake narratives work psychologically.[37] Among the first to use this conspiracy was *Breitbart News* in 2016, becoming a central storyline for Trump during the presidential campaign and in his presidency. The deep state encompasses anyone who preceded Trump in government and is still there opposing him in a make-believe sinister plot to overthrow him. This includes the courts, the Democrats, and liberal academics—all of which are portrayed as espousing left-wing radical ideologies. Trump's election was seen by his base as an opportunity to overturn the deep state and, therefore, investigations against him, such as the Russia one, were inserted into the main plot of the conspiracy theory that claims that the deep state is working behind the scenes to overthrow him and regain power.

Once believed, the conspiracy conditions Trump's followers to filter out news and facts that are contrary to it, seeing them as well as part of the conspiracy. There is no way to dispel the conspiracy since it plays into a deeply held sense in many that the government is always hiding something, so that nothing is as it seems, nor happens by accident. Psychologist Sander van der Linden encapsulates this whole stream of consciousness as follows:[38]

> Conspiracy theorists rarely simply endorse a single conspiracy theory. Rather, belief in one often serves as evidence for belief in others, and this quickly turns into a worldview, i.e., a lens through which we view the world, with new information about world events processed not according to the weight of the evidence but rather in terms of how consistent it is with one's prior convictions. For example, studies have shown that people who believe in conspiracy theories often espouse mutually contradictory explanations about the same event, and are even eager to endorse entirely made-up conspiracy theories. In sum, it's not really about the actual evidence anymore, but rather about whether a theory is consistent with a larger conspiratorial worldview.

Once drawn in, the victim of a conspiracy theory will no longer tend to interpret and judge events in the world in any objective way, but in terms of the insinuations of the theory itself. Those who believe the deep state conspiracy will tend to evaluate any critique against Trump as simply a plan of action by the members of the state.

EPILOGUE

As cultural critic Henry Giroux has observed, the fake news syndrome is emotionally and cognitively destructive because it deeply alters people's understanding of facts and reality, controlling their perceptual filters to the point where the believers of the falsehood will see nothing but what they are told to see by the congeners of the falsehood, leading to the acceptance of an alternative reality, as discussed in chapter 2:[39]

> Trump's language attempts to infantilize, seduce and depoliticize the public through a stream of tweets, interviews and public pronouncements that disregard facts and the truth. Trump's more serious aim is to derail the architectural foundations of truth and evidence in order to construct a false reality and alternative political universe in which there are only competing fictions with the emotional appeal of shock theater.

Giroux's phrase *competing fictions* is an insightful one. It describes what is unfolding in a post-truth era in which science competes with pseudoscience, urban legends with facts, conspiracies with truths, and so on. A perfect example is in the area of climate change, which is a frightening reality that is upon us but is dismissed as a hoax by conspiracy theorists, including Trump. The conspiracy is intended to create confusion, connecting the real science itself to the deep state. It is relevant to note that *The Oxford English Dictionary* named "post-truth" as its 2016 word of the year,[40] just before the American election, referring not only to Trump's mendacious oratory tactics but also to the spread of disinformation through cyberspace, leading to populist movements such as Brexit based on xenophobic conspiracies. In a post-truth world, false beliefs take precedence over logic, science, and reasoning.

Needless to say, not all beliefs are easily manipulated by conspiracies. Some are actually necessary, initially guiding the discovery of facts and can thus be either confirmed or rejected. In this sense, beliefs are part of intuition and inference. Other beliefs are formed instead through rearing that might inculcate falsehoods about the way things are. These do not involve thinking about issues, just reacting to them, accepting even factually impossible things as valid. In Lewis Carroll's

Through the Looking-Glass, the White Queen encapsulates this type of belief system as follows: "Why, sometimes I've believed as many as six impossible things before breakfast."[41]

The post-truth era was predicted by French social critic Jean Baudrillard, who explained it through the notion of the simulacrum (chapter 3), or the idea that fiction and reality have become a cognitive amalgam, much like the state of mind possessed by our ancient ancestors who likely believed their myths to be empirically true, lacking any scientific knowledge to replace them.[42] Baudrillard emphasized that the simulacrum emerges gradually in four stages. First, there is the normal state of consciousness, inhering in a straightforward ability to distinguish between reality and fantasy. This is perverted by constant exposure to fictional portrayals and false beliefs; this is the stage in which Kellyanne Conway's "alternative facts" (chapter 2) and Alex Jones's conspiracies create doubt about reality. This state of mind leads to an Orwellian breakdown between reality and fiction. From this the simulacrum crystallizes. An example that Baudrillard used to illustrate how the simulacrum works was Disney's Fantasyland, which is a simulation of fictional worlds in which visitors can immerse themselves and experience it as more real than real. Eventually, as people engage constantly with the simulacrum, everything—from politics to art—is reduced to a simulacrum whereby truth and falsity blend together unnoticeably.

The 1999 sci-fi movie *The Matrix* portrayed a world in which the simulacrum had become the norm. The main protagonist, Neo, experienced life "on" and "through" the computer screen, and his consciousness was shaped by that intersection. It is relevant to note that the producers of the movie had approached Baudrillard to be a consultant. Seemingly, he turned them down. In the world of the matrix, there is little time and opportunity for reflection on content, given the enormous amount of information plastering our attention on a routine basis, so that we move from idea to idea in a constant flow of information. In this cognitive environment, anything is believable since memory is blocked, and without it reflection can hardly operate. We live in a world where memory is diminished. What happened, literally yesterday, is passé and meaningless. This has led to various mental conditions that are products of the world of the matrix. One is called "source amnesia" by some psychologists and social critics, which refers not so much to a loss of memory but rather to a loss of how and when a memory was created. Susan Greenfield elaborates as follows:[43]

> Memories will now be free-floating, no longer tethered to any personal context. If you have source amnesia, all your memories will blur together instead of being compartmentalized into specific incidents. You may remember a fact but not how and when you learned it. Your recollections would be more like the memories of a small child or a nonhuman animal, hazily aware of order or chronology, and therefore any meaning. Your detailed life story will make no sense, not even to you.

The late Canadian communication theorist, Marshall McLuhan, foresaw danger in the enthusiasm over new media.[44] He warned that they may make us mere "spectators," inclined to abrogate our responsibility to think and act independently, thus debilitating true democracy and meaningful discourse. In a similar vein, MIT linguist and political theorist Noam Chomsky, paraphrasing the American social critic and journalist Walter Lippmann, made the following relevant observation:[45]

> Now there are two "functions" in a democracy: The specialized class, the responsible men, carry out the executive function, which means they do the thinking and planning and understand the common interests. Then, there is the bewildered herd, and they have a function in democracy too. Their function in a democracy, Lippmann said, is to be "spectators," not participants in action. But they have more of a function than that, because it's a democracy. Occasionally they are allowed to lend their weight to one or another member of the specialized class. In other words, they're allowed to say, "We want you to be our leader."

> That's because it's a democracy and not a totalitarian state. That's called an election. But once they've lent their weight to one or another member of the specialized class they're supposed to sink back and become spectators of action, but not participants.

It is truly terrifying to think that in the modern world, truth and falsehood, science and conspiracies, reality and fiction have merged into a simulacrum. To cite Richard Wooley once again, it is truly horrifying that perhaps Trump's most effective tactic has been to erode critical thinking:[46]

> His [Trump's] ideas channel confusion and anger into comfortable solutions that prey on fear but will never address the root causes, which include growing income disparity and diminishing standards in public schools. What can those of us who eschew the illusion of knowledge and seek the truth do? Charles Mackay wrote, "Men, it has been well said, think in herds; it will be seen that they go mad in herds, while they only recover their senses slowly, and one by one." So go find a member of the herd. Appeal to his mind by listening carefully and creating an exchange, and then introduce elements of critical thinking.

5
GASLIGHTING

False words are not only evil in themselves, but they infect the soul with evil.

—Plato (c. 429–347 BCE)

PROLOGUE

The Renaissance philosopher Erasmus once wrote that "Man's mind is so formed that it is far more susceptible to falsehood than to truth."[1] These words encapsulate one of the main themes of this book—namely, that we are all prey to the master liar because he is skilled at manipulating our minds through language that generates obfuscation, ambiguity, and doubt, as well as evoking hidden fears, hatreds, and resentments, in his ingenious scheme to gain trust, support, and backing. His mendacity is made up of potent, unconscious, semantic ingredients that, like a chemical reaction, weaken our ability to think clearly and to reason about things critically. In fact, one of the direct effects that a cunning, manipulative liar can bring about is a state of mind that doubts reality or else accepts the untruths of the liar unwittingly as true, often because they are designed to stoke hidden fears and resentments. The use of dog whistles, of metaphorical language, and dissimulation falls into this category of the Art of the Lie, called colloquially "gaslighting." The intent of the liar in this case is to control the perception of reality through an Orwellian form of verbal artifice that impels people to see in their minds what the liar wants them to see.

The use of dog whistles, rather than direct reference, is a key tactic in the gaslighting strategy. It is, like most of the liar's tactics, a form of doublespeak; in this case the strategy inheres in evoking a referent A by talking about B. For example, Trump's wall metaphor is at one level an allusion to border security (referent B), but at another level it is doublespeak for xenophobia (referent A). This double entendre allows the liar to get away with outrageous things. So, he can say B habitually and then, later, deny indignantly that he was talking about A instead. As Amanda Carpenter notes, this duplicitous tactic started with Nixon and has been perfected by Trump.[2] Not only that, but when called out on his lies, the gaslighter will escalate the denial, doubling and tripling down and rejecting contrary evidence with more false claims. In this way he is constantly sowing doubt and creating mental confusion.

Overall, gaslighting is a tactic intended to avoid problematic referents directly, evoking them instead by innuendo and allusion. The constant use of doublespeak allows the liar to manipulate people's perception of reality by blending referents in subtle ways so that it becomes difficult to distinguish one from the other. The goal is to make interlocutors second-guess his choice of words, thus enhancing the possibility of making them dependent on him and him alone to understand what they mean. Bryant Welch argues that gaslighting has been a political strategy in American politics for decades, arising from the mindset produced by the mass marketing and advertising techniques in which we are submerged: "Gaslighting comes directly from blending modern communications, marketing, and advertising techniques with long-standing methods of propaganda. They were simply waiting to be discovered by those with sufficient ambition and psychological makeup to use them."[3]

Gaslighting has many sides to it, and only some of them can be discussed in this chapter. But in all its manifestations, the overriding characteristic is the artful deployment of innuendo and indirect reference to beliefs and concepts that cannot be articulated overtly for fear of reprobation or reprisal. This is accomplished primarily through the use of dog whistles that indirectly legitimize racist or xenophobic beliefs by alluding to them in a coded way. The language of dog whistles is never directly referential—it produces its gaslighting effects through evocation, thus duping people into believing that they themselves have formed the thoughts that the liar stokes gradually and repetitively. The victims of the master liar rarely realize that they are being manipulated, accepting his dog whistles as part of the ongoing attack against perceived enemies.

Gaslighting is dissimulation at its Machiavellian best, allowing the master liar to ally himself to a particular group by feigning to adopt their goals, as Trump has accomplished with the white evangelicals in the United States (as discussed), without any obligation to actually be a card-carrying member of the group. Dissimulation is not exclusively a gaslighting tactic, of course. Nevertheless, it is an intrinsic one for producing gaslighting effects. Above all else, it allows the liar to avoid accountability, since the meaning of his words cannot be pinned down to a definite content. Slippery, duplicitous language of this kind shields the liar from any direct accusation while it keeps him firmly ensconced in the good graces of his followers.

THE GASLIGHTER

As mentioned in chapter 1, Oscar Wilde's metaphor of the fog has been adopted here to describe the kind of doubts and feelings of uncertainty that the masterful liar can produce in us, projecting us into a mind fog where nothing is for certain. This is a perfect metaphor to describe gaslighting. As discussed throughout this book, the skillful liar knows how to produce mental confusion by using words and phrases that create vague images in the mind that are intended to stoke buried resentments and even hate by innuendo. In effect, the liar knows how to create ideas in the mind fog that cannot be pinned down directly, but which nevertheless convey latent meanings through allusion and innuendo. To reiterate, he refers to A by talking about B. There is no empirical way to demonstrate that A is the intended content, and this is what makes such doublespeak powerful. It projects people into a mind fog via the vehicle of suggestive reference.

The term *gaslighting* comes from a 1938 play, *Gas Light*, by Patrick Hamilton, which was adapted into the 1944 movie *Gaslight*, directed by George Cukor,[4] in which a man manipulates his wife to the point of exasperation whereby she starts to believe that she is losing her mind. Amanda Carpenter's perceptive (and chilling) book, *Gaslighting America: Why We Love It When Trump Lies to Us*, looks at how Trump, like the character in the movie, has been manipulating his followers through deceptive language based on innuendoes that assail the level of reason, destroying its operation and projecting people into a dark corner of the mind where pent-up resentments and fears are stoked and legitimized.[5] This is a sinister Machiavellian strategy, and one of the most dangerous mind-twisting tactics of the fox, as Machiavelli described the liar-prince, since the content that he wishes to convey is not done so by direct communication, but as coded language that spreads among followers like a secretive military cryptography, mobilizing them to act out their inner resentments and beliefs through actions and behaviors that would have been previously impossible and even unthinkable. The reach of this gaslighting code has become extensive in the internet age, since it now gains momentum through memes and viral videos. This makes us all the victims of gaslighting, since even those who detect its cryptic intent are powerless to counteract it. This is why it has been used efficaciously by dictators, narcissists, and cult leaders, among others of similar ilk. It works best when it is carried out methodically and with repetitive timing.

Dog whistles exemplify how gaslighting works, as Karen Grigsby Bates has remarked in an interesting editorial.[6] One example is Trump's reference to countries governed by Africans, such as Haiti, as "shithole" countries, which he is purported to have said during a cabinet meeting.[7] When confronted with this racist statement by reporters, he denied using the term, saying that he "Never said anything derogatory about Haitians other than Haiti is, obviously, a very poor and troubled country."[8] The dog whistle was not the term *shithole countries* per se—that was an overt racist statement—but in the follow-up "very poor and troubled country" phrase, which alludes to poverty and trouble as inherent characteristics of African societies. Trump is a master in using this kind of dog whistle. For instance, his use of "criminals" and "rapists" in reference to "some" Mexican immigrants, as discussed several times, are xenophobic dog whistles referring to the "bad hombres" who live in Mexico. Painting them as delinquents allows Trump to allude to their place of origin as socially inferior.

This kind of language is nefariously effective because it has plausible suggestibility. So, while we might recognize "very poor and troubled country" as a dog whistle, we might also see in it a "grain of truth." This is what makes dog whistling so slippery and dangerous—its main suggestion is racist, but it is also seen as referring to something plausible in a stereotypical way. Stereotyping of this kind is designed to typecast people in a skewed way by alluding spuriously to perceived traits as characteristic of a people as a whole. It plays on false generalizations by twisting plausibility to generate prejudice. Typecasting groups into categories (bad hombres, troubled, etc.) allows Trump to attack them through abstractions and allusions, rather than directly.

Trump's followers and allies do not perceive his dog whistles as blatant strategies of stereotyping. Living in the doublethink mind fog, they tend to perceive them primarily as part of a general clarion call to arms to overturn the deep state. They are thus employed as subtle verbal weapons in the cultural war that the liar-prince creates and then continues to stoke through them. There is likely nothing he could say or do that would erode support from his followers, as long as it is in the name of the greater cause of taking down the political enemies that he and his followers see as the source of their troubles. Carpenter identifies five distinct gaslighting strategies, which she calls "stake a claim," "advance and deny," "create suspense" by announcing forthcoming evidence, "discredit the opponent" with personal attacks, and "win" by self-proclamation. Many of these have been discussed under different rubrics in this book. But what differentiates gaslighting to some extent from other forms of lying is that it conditions the mind into seeing things that are not there, but only thought to be there. It projects people into a "Twilight Zone" of vague thoughts that are felt to have great import, and which are understood by those in the know. Gaslighting discourse begs the question "Is this what he means?" If the answer is "Yes," it is then virtually impossible to exit the Twilight Zone, since it would entail bitterly accepting the truth that the others are not enemies of the state—just a state that is diverse and multilayered.

Carpenter's list can be seen to operate in all manifestations of dog whistling. Consider the "bad hombres" one. With this statement Trump has staked a racist claim—Mexicans are bad people. However, he can always deny that he meant this by simply saying that only "some" are bad, thus putting forth a strategic form of denial that does not actually deny anything. The pronoun *some* is an indefinite one, leaving it to interlocutors to figure out the extent of the badness among Mexicans. So, what is he going to do about it? As he said throughout the campaign, he is going to build a wall at the southern border and "Mexico will pay for it." When attacked for not coming through with the latter part of this promise, he called his plan a "win," because the wall will be built even if Mexico will not pay for it directly, but in other vague ways—a masterful gaslighting stroke. Of course, all of this occurs in the mind fog, and this is what makes gaslighting so effective, denying clarity to words and their meanings.

William Safire has provided an insightful characterization of dog whistling, defining it as the strategic use of subtle wording, whereby some people hear something in the wording that others do not.[9] Researcher Amanda Lohrey found that dog whistling is an effective political strategy, designed to appeal to the greatest number of voters without necessarily alienating others.[10] The use of "family" and "values" during elections are actually dog whistles, which resonate with conservative Christian voters, without seeming to be moralizing speech that would push secular voters away.

Gaslighting involves above all else a shrewd use of metaphor. Consider the deep state metaphor again (chapter 4)—a trope that alludes to corruption in government that can only be combatted by someone (Trump himself) capable of "destroying" this hidden state by "draining the swamp" in which it purportedly exists. Metaphorical language makes seemingly disparate referents coalesce seamlessly into an overall scenario. So, the deep state metaphor achieves several things at once—it taps into a belief that liberalism and its elitist politically correct discourse has ensconced itself "deeply" into American politics and society at large and thus needs to be eradicated, at the same time that it fits in with the conspiratorial narrative of persecution that Trump is spreading to protect himself—persecution from the political left. Trump has repeated this trope so many times, in public and in tweets, that it now has become a colloquialism, hiding in its subtext a racial dog whistle—namely, that the deep state is secretly supporting the previous president, Barack Obama. It is thus both a conspiracy theory in itself and a gaslighting technique that encodes a subtle dog whistle that is intended for, and understood by, a particular group of people, with no requirement to show that such a state exists *in reality*.

The fact that many believe such metaphors as *real* is evidence that we do not perceive reality directly but through the lens of linguistic depictions. When a metaphor is devised to conjure up false referents, a weird thing happens—we know they do not exist in reality but we still believe that they hide some truth, arguably because we feel that they contain messages that need to be decoded. There is no real thing called a *unicorn*. But the word still conjures up the image of a horse with a single straight horn jutting out from its forehead. We get this image from mythic stories, of course. But the fact that it pops up in the mind via the word impels us to accept it as having some hidden significance, even though we know it is an imaginary referent. Actually, all language works this way—it produces a sense that the world can be codified and stored in the mind through words. In the hands of the liar-prince, language can be manipulated to evoke plausible worlds with no requirement to prove their validity or even existence.

So, a metaphor such as the deep state one, which does not point to anything specifically, has plausibility, when repeated over and over, like the unicorn referent. It refers to a set of references by allusion, making it virtually foolproof from counterattacks, since there is no need to demonstrate its reality in the same way that there is no need to prove or disprove the existence of a unicorn. This type of verbal strategy is what makes gaslighting so dangerous: allies of the liar understand the coded allusive meanings—referential scenario A—and will even expand on them, manufacture ersatz evidence for them, creating false narratives that are entirely based on them—metaphorical referents B. Opponents are thus at a loss to counter these verbal narcotics because it is virtually impossible to rationally attack them. To attack A, for instance, would allow the liar to say that B was intended literally; vice versa to attack B as a dog whistle would allow the liar to similarly claim that he meant something else.

There is no effective verbal antidote to this type of Orwellian doublespeak. It allows the master liar to render his language impervious to counterattacks and criticism. He can even go back on his words with no adverse consequences. An example of this occurred when Trump reversed a campaign pledge, declining to call China a "currency manipulator" after being elected, as he had done constantly during the presidential campaign, claiming that China had *now* stopped the "bad behavior" and attributing this epiphany to his own skills as a deal maker and political leader. He claimed that this would lead to more jobs and economic prosperity. When it did not, he turned against China by imposing tariffs, with not a whisper of condemnation from his followers. As Machiavelli knew, changing one's mind is allowed if it undergirds the achievement of something tangible, including any gifts he hands to them (such as jobs, judges, etc.). As the Italian shrewd political adviser suggests in his manifesto (and as the Erasmus citation implies at the start of this chapter), people are much too easily deceived by words and gifts:[11]

> For, as I have said already, the ambitious citizen in a commonwealth seeks at the outset to secure himself against injury, not only at the hands of private persons, but also of the magistrates; to effect which he endeavours to gain himself friends. These he obtains by means honourable in appearance, either by supplying them with money or protecting them against the powerful. And because such conduct seems praiseworthy, everyone is readily deceived by it, and consequently no remedy is applied.

A parallel use of this strategy—promising prosperity with no real evidence that it will occur—can be seen in the following statement by Mussolini:[12]

> Fascism establishes the real equality of individuals before the nation. The object of the regime in the economic field is to ensure higher social justice for the whole of the Italian people. What does social justice mean? It means work guaranteed, fair wages, decent homes, it means the possibility of continuous evolution and improvement. Nor is this enough. It means that the workers must enter more and more intimately into the productive process and share its necessary discipline. As the past century was the century of capitalist power, the twentieth century is the century of power and glory of labour.

It is not coincidental that Mussolini uses the same kind of metaphorical superlatives as does Trump: "real," "higher," "power and glory." These are designed to evince images of greatness and to suggest that he, the Duce, will guarantee prosperity. Mussolini's metaphorical doublespeak gaslights his followers with equivalencies between "the economic field" and "social justice," between fascism and the "glory of labour." The overall intent is to promise economic gifts ("fair wages," "decent homes") through the "possibility of continuous evolution and improvement." What does this really mean? As in any gaslighting tactic, the meaning is left nebulous, suggesting indirectly that "evolution" and "improvement" are what fascism will ensure, without concretely asserting how such an ideal society will be realized.

Mussolini was adept at coining catchy patriotic slogans, imbuing them with shrewd innuendoes and dog whistles. The vague ideas that they referenced resonated with the sense of "nationalism" that was coagulating in Italy at the time, becoming symbolic of the new world order that Mussolini was proclaiming. A few examples of his metaphorical savvy will suffice:[13]

Giovinezza ("Youth"). This was Mussolini's slogan meant to draw young people into the fascist fold, since young people were the ones whose support he would need to move forward politically; significantly, it became a powerful metaphor supporting fascism that was uttered throughout society, effectively neutralizing opposition by connecting youth, the rebirth of the true Italy, and fascism into a symbolic fusion.

Italia Imperiale ("Imperial Italy"). The fascists used this moniker as a rallying cry to generate support for their imperialist plan to gain dominion over the Mediterranean area, recalling the glory days of the Roman Empire.

Italia Irredenta ("Unredeemed Italy"). This catchphrase advocated the importance of "redeeming" the former Italian-held territories outside of Italy, thus further reinforcing the plan of the fascist regime to regain the past glory of the Roman Empire.

Italianità ("Italianness"). This term was actually coined during the *Risorgimento*, the period of the unification of Italy in the nineteenth century; it was recycled strategically to stress the importance of Italians coming together under one regime—fascism.

Mare Nostrum ("Our Sea"). This was actually used by the ancient Romans in reference to the Mediterranean Sea as part of their empire. Mussolini employed it to impart a sense of the importance of regaining control of the area under fascism.

It is unlikely that Mussolini would have attracted so many zealous acolytes without such language. The dog whistles and slogans used by Trump are eerily similar if not identical to the ones used by Mussolini. Trump even called himself a "nationalist" at a rally in Houston on October 22, 2018:[14]

> You know, they have a word—it's sort of became old-fashioned—it's called a nationalist. And I say, really, we're not supposed to use that word. You know what I am? I'm a nationalist, okay? I'm a nationalist. Nationalist. Nothing wrong. Use that word. Use that word.

In typical Machiavellian repartee, after being attacked for using this racist dog whistle, Trump claimed that he was unaware that the term *nationalist* carried any racist connotations, defending his use of the word as an act of restoring its proper sense of "patriotism." It is not surprising that virtually no one within his fold condemned him for using this dog whistle, which was used constantly by Mussolini and Hitler. His response was effective doublespeak because it challenged his counterattackers to defend the meaning of the word *nationalist* as a dog whistle, which put the challengers on the defensive. In effect, he used the same type of duplicitous gaslighting strategy—saying B to mean A and then denying any knowledge of A's real meaning.

Gaslighting eventually leads victims to disbelieve anything other than what the gaslighter tells them is true (even if it is demonstrably false). As Bobby Azarian remarks, the gaslighter knows how to make people suspicious of anything others might say or do:[15]

> The president knows that with an uninformed group of people, it's his word against the "fake news media." If his followers did become

cognizant of gaslighting as a political tactic, he'd likely just flip the script by telling them that it is the journalists, pundits, and intellectuals who are trying to gaslight them. While this might sound absurd to some, to those with little education and mental vulnerabilities, the confusion can shake their confidence, sowing seeds of doubt that can set them down the path of questioning their entire reality.

An implicit principle of gaslighting is to make sure that the lie is always colossal (not a simple white lie) so that people will be inclined to accept it as true. In other words, the more outrageous the lie, the more people are likely to believe it. Hitler called this the "big lie" technique:[16]

> In the big lie there is always a certain force of credibility; because the broad masses of a nation are always more easily corrupted in the deeper strata of their emotional nature than consciously or voluntarily; and thus in the primitive simplicity of their minds they more readily fall victims to the big lie than the small lie, since they themselves often tell small lies in little matters but would be ashamed to resort to large-scale falsehoods. It would never come into their heads to fabricate colossal untruths, and they would not believe that others could have the impudence to distort the truth so infamously. Even though the facts which prove this to be so may be brought clearly to their minds, they will still doubt and waver and will continue to think that there may be some other explanation. For the grossly impudent lie always leaves traces behind it, even after it has been nailed down, a fact which is known to all expert liars in this world and to all who conspire together in the art of lying.

Hitler was himself a master of the big lie (recall the Aryan myth), never allowing his followers to forget it, and never admitting fault or conceding to the counterattacker.[17] The deep state and birther conspiracies, among many others, are Trump's big lies, and this is why, arguably, that they produce the gaslighting effects that they do.

ARTIFICE

Gaslighting is verbal artifice at its most sinister, because it involves the Orwellian ability to generate images in the mind that are ambiguous. As discussed several times, behind this kind of artifice is deceptive metaphor. Consider the deep state metaphor one more time. The term *state* does not refer to government per se but to a world composed of "thought police," while *deep* implies something secret or conspiratorial. The totalitarian images that this metaphor evokes allow the liar to perpetrate his conspiracy effortlessly through the images themselves, with no requirement to verify them as real. Those in the deep state can now be depicted as the "enemies" of democracy and freedom. It is a brilliant ploy, reversing the tables on his attackers who claim that he, Trump, is the real enemy of democracy. It allows him to associate liberalism with Soviet-type totalitarianism, run by Leninist-type elitists who impose political correctness as their main weapon of thought control.

Metaphorical artifice works psychologically because it impels people to perceive hidden connections among referents in terms of image schemas, as linguists point out.[18] That is to say, the deep state image is crafted in such a way that allows Trump to subtly evoke a portrait of liberals as autocrats who are running America behind the scenes. This implies a call to action—namely, the deep state must be defeated. In other words, the metaphor produces semantic tentacles that allow Trump to interconnect elements in his false narrative cleverly, as can be seen in the following tweet, in which he connects the deep state directly with the political left and the fake news media, suggesting at the same time that his rise to power has already made a dent in the supposed cabal, leading to significant accomplishments (jobs, appointment of conservative justices, and so on):

> The Deep State and the Left, and their vehicle, the Fake News Media, are going Crazy—& they don't know what to do. The Economy is

> booming like never before, Jobs are at Historic Highs, soon TWO Supreme Court Justices & maybe Declassification to find Additional Corruption. Wow![19]

Given the obvious power of such verbal artifice, it is informative to take a brief historical digression into the scientific study of metaphor, which started in the first decades of the twentieth century, leading to a significant 1936 book by literary critic and educator I. A. Richards, *The Philosophy of Rhetoric*.[20] Richards's treatise triggered a paradigm shift in the study of language. In it, he argued that metaphor could hardly be classified as a replacement of literal meaning for decorative or stylistic purposes, but rather that it was basic to the way we interpret the meanings of words as they are used in specific communicative contexts.[21] Without going into the relevant details of Richards's argument here, suffice it to say that by mid-twentieth century, it became obvious that a specific metaphor, such as "deep state," is not just colorful language—it reveals how we blend concepts together to generate suggestive ideas. A watershed 1977 study showed, in fact, that metaphors pervade common, everyday speech. Titled *Psychology and the Poetics of Growth: Figurative Language in Psychology, Psychotherapy, and Education*, it found that speakers of English uttered, on average, a surprising three thousand novel metaphors and seven thousand idioms per week.[22] It became saliently obvious that metaphors could hardly be characterized as a deviation from literal speech, or a mere stylistic accessory to literal conversation. A groundbreaking 1980 book by George Lakoff and Mark Johnson, *Metaphors We Live By*, put the finishing touches on the theory that metaphors shaped thought and thus could be used to manipulate it.[23] Their analysis allows us to understand why the metaphors of a Mussolini or a Trump are so persuasive.

Lakoff and Johnson showed how we use metaphor to convert concrete experiences into abstractions, permitting us to see things in the mind as if they were physically real. From these, we can draw inferences and entertain plans of action. Since the deep state metaphor evokes the image of a group of people conspiring secretly together, which has "deep roots" and thus is difficult to eradicate, it suggests that an eradicator, such as Trump, is needed to eliminate it. All of this unfolds in the mind, but it suggests external action. In fact, Trump is convinced that his followers will take even violent action to protect him, issuing confidently in a March interview with *Breitbart News* the warning that he cannot be defeated, since he has the support of the police, the military, and "bikers for Trump," who, if they have to, will get very tough on opponents, and that this would be "very bad."[24]

Consider another of his duplicitous metaphors, which he used throughout the campaign—namely, his pledge to build a "wall" to keep Mexicans and other illegal immigrants out of the United States. The wall is verbal artifice at its best, because it is much more than a physical barrier, as already mentioned; it is a metaphor that can be twisted to mean anything he wants, such as providing "safety" and "security":

> We need the Wall for the safety and security of our country. We need the Wall to help stop the massive inflow of drugs from Mexico, now rated the number one most dangerous country in the world. If there is no Wall, there is no Deal![25]

The barrier image schema is psychologically effective because it suggests a course of political action—it is needed to impede foreign "invaders" from coming into the country who will lay it waste with their criminality. It is an image schema that is grasped tangibly. It is also subtly allusive to the Great Wall of China, a fortified wall in northern China, extending some 1,500 miles from Kansu province to the Yellow Sea north of Beijing, and first built around 210 BCE as a protection against invaders. As one of Trump's staunch supporters, Senator Lindsey Graham bluntly admitted in late December 2018, speaking to reporters in front of the White House: "The wall has become a metaphor for border security."[26] This can be read in various ways, evoking an actual physical barrier but also a cultural barrier that aims to clamp down on the "invasion" of diversity into the United States.

Although he did not name this type of verbal artifice as metaphor, as has been done here, Machiavelli was well aware of the potency of this kind of language to shape beliefs and to evoke emotions that can be easily manipulated to spur on political action. At a literal level, the message is seemingly a straightforward one: "The wall is needed for security." There is no argument against this. But at a metaphorical or coded level, the wall reveals another message. First, it implies blockage of the inflow of illegalities. Trump has used it to claim that it is the main solution to an ever-expanding national drug abuse problem. This is misleading, since most drugs enter the country not through the southern border but through various means (by airplane) and other locales (ports of entry). But by blaming Mexico, the wall metaphor fits in with his conspiracy theory of what is "wrong" with America and how to restore it to its grandeur, recalling the similar style of metaphorical artifice used by Mussolini (above). The wall thus is a multilayered metaphor that works at different levels of mind and emotion.

In the end, the wall metaphor is a "Just So Story"—a term taken from *Just So Stories for Little Children* (1902) by Rudyard Kipling in which made-up stories are concocted to explain animal features and appearances, such as the origin of the spots on leopards.[27] A "Just So Story" is an artfully contrived metaphor, having no basis in fact. It is certainly an apt moniker for metaphors such as the "deep state" and "wall" ones, which are, essentially, Just So Stories.

To maximize the mind fog produced by such stories, the skilled liar must know how to deliver them effectively to his followers, as discussed previously. Artifice involves both verbal and nonverbal communication tactics. In the same way that an expert comedian knows how to deliver a joke to maximum effect, so too the gaslighter must ensure that his delivery evokes the anticipated reaction. This is arguably why at his rallies, while he is discussing a certain situation, out of nowhere Trump comes up with a "punch line" that grabs everyone's attention to great effect. For example, during a speech in Alabama in 2017 in support of Senator Luther Strange, out of the blue Trump made the following statement:[28]

> Wouldn't you love to see one of these owners, when somebody disrespects our flag, to say, "Get that son of a bitch off the field right now? Out! He's fired. He's fired!"

The "joke" evoked a roar of approving laughter from his audience, not only because it was unexpected, but also because it was a dog whistle that resonated with many. The comedic effect was reinforced by the sequential structure of the delivery: the framing ("Wouldn't you love to see one of these owners . . .") followed by the bone of contention ("when somebody disrespects our flag, to say, 'Get that son of a bitch off the field right now'?"), ending with the punch line ("Out! He's fired. He's fired!"). This dog whistle does not appear to be overtly racist, since it is presented as a plea for patriotism, symbolized by the ritual standing for the national anthem at public events, such as at a sports event, as a way to show respect for the military. But there is little doubt that it was aimed at African Americans, since the protest was started by a football player to protest the mistreatment of African Americans by the police. So, again he used B (patriotism) to imply A (racial politics and identity), a classic doublespeak ploy.

This type of verbal artifice—pretend to say something, but intend something else—is sometimes called the art of dissimulation. As French statesman Cardinal de Richelieu so perceptively remarked: "To know how to dissimulate is the knowledge of kings."[29] Hannah Arendt has argued that this kind of tactic is extremely dangerous, because it allows the liar to generate the perception of reality he needs to gain power:[30]

> Before mass leaders seize the power to fit reality to their lies, their propaganda is marked by its extreme contempt for facts as such, for

in their opinion fact depends entirely on the power of man who can fabricate it.

As discussed several times, the same type of verbal artifice was used by Trump with his birther conspiracy theory. It was evasive—Trump claimed that he was merely looking to uncover the "truth" about Obama's identity—and an indirect reference to Obama's race. Because of this duplicity, Trump could always slither out of any accusation of racism against him because he would simply say that his aim was to get to the "truth" of the matter.

Another well-known example of a racist dog whistle occurred in a thirteen-minute-long 2015 interview conducted by CNN when Trump was questioned about his proposal to ban Muslims from entering the United States. His answer was clearly intended to evoke images of 9/11: "Most Muslims, like most everything, I mean, these are fabulous people . . . but we certainly do have a problem, I mean, you have a problem throughout the world . . . It wasn't people from Sweden that blew up the World Trade Center."[31] There is absolutely no doubt that he was opportunistically suggesting that Muslims are a threat to America.

THE GREAT PRETENDER

As Machiavelli wrote, the successful prince must be a "great pretender and dissembler."[32] The skill of pretending or feigning to endorse a cause is one of the liar-prince's most strategic abilities—recall the discussion on how Trump and Mussolini espoused religious causes, even though there is no evidence that they were religious before rising to power. Machiavelli saw this ability to pretend and dissimulate as the vulpine qualities that a liar-prince must possess, allowing him to feign rectitude and adopt moral causes (even though he might not believe in any of them).

Trump understood from the outset of his presidential campaign that the ultimate battleground in the culture wars that had already come to the surface before him was the fervent support that the ultraconservative religious segment of society would give to him if he espoused their causes. Trump has consistently played to the sense of moral panic that the evangelicals have felt in an era of secularism and political correctness, supporting their beliefs and agendas in explicit ways. White evangelical groups in the United States in particular are, as the Machiavellian fox understood, among the most vociferous and tenacious leaders in America's culture war, firmly believing that America's religious foundations were eroded by secularism, relativism, and the acceptance of all kinds of non-Christian faiths that have been seeped into America as equal. Because of their unwavering, impassioned, and fervid devotion to their beliefs, which unites them politically, they have had considerable impact on election outcomes, especially since they have also become skillful users of social media to promote their views broadly and to unite themselves in their common cause.

Again, it is entrenched belief that opens up an individual or group to manipulation by the dissembler, who can easily appeal to believers by taking up their moral cause against what they see as the spread of heathenism. By catapulting himself into a style of political leadership that espouses religious revivalism, Trump knows that this huge slice of society will always protect him against the nonbelievers, so that no matter what he does he is seen as their valiant, moral warrior who will restore righteousness and bring about a retrieval of moral rectitude in society at large. With clever verbal artifice, Trump comes across as authentic to his religious followers, assuring him the allegiance that is given to preachers and religious leaders. When Trump was interviewed during the campaign about his views on abortion, he insinuated that those who underwent abortion must be "punished" in some way[33]—a term that is hardly presidential in the traditional sense, but rather something that a preacher would say.

At the 2019 National Prayer Breakfast, Trump, like a true Machiavellian fox, proclaimed to his religious audience that he "will never let you down; I can say that. Never."[34] He also recalled his supposed opposition to abortion with a sermonlike statement: "As part of our commitment to building a just and loving society, we must build a culture that cherishes the dignity and sanctity of innocent human life. All children born and unborn are made in the holy image of God."[35] This is a remarkable statement in the light of his mistreatment of immigrant children at the southern border. Any despicable actions and behaviors that he might manifest are seen in a cultural light; what counts is the final moral apocalypse that the leader is supporting. In an insightful *Guardian* article, journalist Julian Borger has written that evangelicals have put aside Trump's obvious "sinful life" and accepted him as their leader because they see him as a modern-day King Cyrus, the sixth-century Persian king who liberated the Jews from the captivity of the Babylonians.[36] Cyrus was a nonbeliever who was seen as the vessel of God so that the faithful could be liberated and their religion restored. Trump's sins and his attack on democratic values, such as his constant assault on freedom of the press, are thus forgiven because he is viewed as the leader who will do whatever is necessary, as God's vessel, to liberate society from the claws of immorality and secularism.

The fox thus portrays himself as a lion, pretending to be a valiant warrior for morality. Any attacks against him allow him to portray himself as a victim of the secular deep state, thus gaining protective support from the followers—the lion cannot fight the battle on his own; he needs an army behind and in front of him. So, he pretends to be on the side of religious truth. Trump emerges as a righteous warrior who has taken on the plight of all the victims of the deep state unto himself, becoming himself a victim of that corrupt state. The fox knows that his lies will be perceived as necessary tactics in the larger battle—they are shields and weapons that he uses to carry out the war. Not surprisingly, Trump's pseudo-religious discourse is cluttered with metaphors that resonate with his followers, including his attack on the liberals in the deep state as "monsters," "evil," "inhuman," "beasts," "savages," and the like. These words are confirmation to his fierce religious followers that he is committed to the cause. The fox in this case comes dressed as a knightly lion.

The same pseudo-religious tactic has been used by previous dictators and even the Mafia, as was depicted realistically in the 1990 film *The Godfather, Part III*.[37] The movie revolves around the Mafia's historical involvement with, and connection to, the Catholic Church, bringing out the fact that the Mafia sees itself, or at least portrays itself, as a quasi-religious institution. At a social event in the movie, the character Michael Corleone is seen donating $100 million "to the poor of Sicily," which he gives to the Church to distribute equitably. The duplicitous implication is that the Mafia is an honorable and charitable organization that traces its own roots to groups of valiant, moral men whose intention has always been to help the poor. Religion was also a frequent theme in the HBO series *The Sopranos*. In episode nine of the second season, for example, a mobster is shot and pronounced dead for approximately one minute, during which time he has a chilling vision of hell.[38] This makes another mobster nervous, because he is very superstitious. So he goes to a priest asking him if donations to the Church would be enough to be forgiven for a life of brutal crime so that he can escape the fires of damnation. Mobsters may or may not have a conscience, but they certainly do understand that their actions are profoundly evil. And like everyone else, they fear retribution.

Mussolini aligned himself from the outset with the Catholic Church, as David Kertzer has cogently argued in his discerning book, *The Pope and Mussolini: The Secret History of Pius XI and the Rise of Fascism in Europe*.[39] Those critical of Mussolini were cast as godless liberals—a situation that is paralleled in Trump's evangelical America. The Vatican and the fascist regime had many differences, but they shared the goal of "restoring" morality—in the former it was a historically valid goal; in the case of the latter it was a pretense. As Mussolini so intrepidly claimed: "Fascism is a religion. The twentieth century will be known in history as the century of Fascism."[40] In such a pseudo-religious scenario, democracy itself is seen as susceptible to moral decay because it allows different voices, religious and secular, to shape its ethical constitution. As a consequence, democracy's open worldview must be debunked as morally destructive, a point articulated tersely yet eloquently by Orwell as follows: "One of the easiest pastimes in the world is debunking Democracy."[41]

By adopting an antidemocracy stance in the name of morality, the Machiavellian fox can thus turn his politics into a Holy War against anyone who does not support him. The scary thought that this evokes is encapsulated in something that American novelist Sinclair Lewis purportedly wrote in the 1930s: "When Fascism comes to America, it will be wrapped in the flag and carrying a cross."[42] Trump's flag-hugging performances, his adoption of evangelism, and his warnings of potential violence are clearly reflective of Lewis's warning.

As any other Machiavellian fox, Trump understood from the beginning that cultural diversity, relativism, and secularism are perceived as causing irreparable moral harm by religious groups to the foundations of society. This has allowed Trump to equate moral corruption with liberal democracy, making it possible for him to attack democratic institutions as morally corrupt. The "enemies of God" are thus liberal judges, secular public schools, academia, left-wing intellectuals, Democrats, and anyone else who sees diversity as a principle of democracy. The threat by such a worldview and politics to the moral fiber of America is seen in apocalyptic terms. In the battle against the forces of evil secularism, Trump emerges as a King Cyrus who will liberate American society from captivity by the liberal deep state.

While pretending to endorse causes is not, technically speaking, gaslighting, it can induce false beliefs in victims in a similar way. Believing that Trump is King Cyrus is illusory, yet somehow it resonates as real to religious supporters. Trump's pretending act is illusion, akin to the illusion of magicians. As magician Ben Chapman states: "For magicians, this means they must perform effects in which people want to believe."[43] For Trump, it means saying things and carrying out actions that religious people want to believe as authentic. The underlying principle of this strategy can be encapsulated as follows: "Tell people what they want to hear and they will believe you."

EPILOGUE

Gaslighting is effective because the liar can play the role of the victim, gaining sympathy and support. The real victims are those who fall prey to the fox's gaslighting strategies. The liar knows that in order to be believed he can never stray from his rhetoric of pretense—as Trump demonstrated in his speech to the audience at the National Prayer meeting. The gaslighter manufactures consent, to use Herman and Chomsky's phrase once again (chapter 4), by carving out an Orwellian mind world where reasoning is overtaken by belief. The ultimate objective of the gaslighter is to control people. With an incessant stream of lies and deceptions, the gaslighter keeps victims in a constant state of insecurity, doubt, and fear. This allows him to exploit them at will, for personal gain.

Because Trump is perceived as a savior-warrior, his rallying cries become those of audiences; his transgressions are seen as necessary to bring about change; his dog whistles are perceived as central weapons in the cultural war. It is somewhat ironic to observe that Trump portrays his deceptive cultural war as a counterculture one, as did the hippies in the 1960s and 1970s, portraying the government as the "establishment" and the liberal democratic state—again quite ironically—as the enemy of freedom and true American values. This may be the most effective of all gaslighting strategies. It can be encapsulated as follows. The establishment (the deep state) has been involved in mind control through political correctness and thus has attacked the freedom of thought on which America was founded. There is thus a need to overthrow the establishment through a revolution that is an upfront one. It is truly sardonic to witness that some of Trump's followers had grown up during the real counterculture era. The hippie revolution instilled in America a sense that things had to change. As typically happens in mass movements, there were bound to be ups and downs. Indeed, near the end of the era, America reelected Richard Nixon with one of the biggest majorities in American history—a veritable setback (despite Watergate shortly thereafter) for the changes the hippies wanted to bring about to society. And the election of Donald Trump as president in 2016 also stands in stark contrast to the hippie goals. But the clever fox evokes the same kind of revolutionary sentiments through a discourse that attacks political correctness as an evil to be eradicated. It is a brilliant strategy indeed.

To conclude this chapter, the gaslighter is an Orwellian Big Brother, who will pounce on any opportunity to inculcate his fabricated images of reality into vulnerable minds, controlling them not through physical threats, but through dog whistles and subtle metaphors. People will follow the liar-prince fervently, if he can make them believe him that he espouses their causes. Machiavelli put it as follows: "Men are quick to change rulers when they imagine they can improve their lot."[44] Machiavelli's Florence was immersed in corrupt politics. He witnessed how clever politicians were able to gain power by employing the Art of the Lie. From this situation, he wrote *The Prince.* Hopefully, America can go beyond the Machiavellian view of successful politics as based on gaslighting and dissimulation.

6
VERBAL WEAPONRY

It is not the lie that passeth through the mind, but the lie that sinketh in, and settleth in it, that doth the hurt.
—Francis Bacon (1561–1626)

PROLOGUE

When attacked, the best strategy, as Machiavelli emphasized, is to attack back or even attack first, preempting the ability of opponents to be effective by putting them on the defensive. "Head them off at the pass" was a cinematic cliché used in Hollywood cowboy movies, which encapsulates the intent of this game plan perfectly. A master liar must anticipate adversity all the time, and be constantly prepared to head off an opponent at the pass by leveling the same kind of attack preemptively on the attacker, thus obverting attention away from himself and putting it on the person who is coming at him. In military parlance, it is called, in fact, a preemptive attack. Machiavelli compares the strategy to a physician treating a malady before it has time to grow and worsen.[1]

In line with military parlance, this kind of tactic can be called "verbal weaponry." The main weapons are deception, denial, and deflection. Their utilization can be seen in several combative gambits: blame the blamer, deny any wrongdoing, deflect attention away from oneself, call one's attackers names that will vilify them, and deflect attention away from oneself by casting doubt on the actions of others. Trump is a master at this type of military verbiage. He will blame anyone who attacks him as being guilty of the same crime of which he is accused or else conceal the truth by constant denial. For example, he accused special prosecutor Robert Mueller in the Russia investigation that was established in 2017 to determine if he had colluded with Russia to win the election, for doing exactly what he himself (Trump) was suspected of doing—lying—in order to insinuate that Mueller's investigation was tainted by political motivations and was thus a political witch hunt. This deflected attention away from the purpose of the investigation and cast suspicion on the investigators themselves. By repeating it over and over, his cross-accusations gained plausibility among his followers, who saw the investigation as verification of the deep state conspiracy theory.

A major weapon in the liar's arsenal is name-calling, which is intended to hurt someone's reputation, good name, or character. It is Trump's main intimidation tactic, which he employs to weaken his opponents and critics, deflecting critical attention away from himself.

Deflections, preemptive lies, deceit, hateful nicknames, and the strategy called "whataboutism" are some of the verbal weapons and shields in the Machiavellian Art of the Lie. They constitute a powerful arsenal of mendacity and dissimulation that the master liar can use to discredit opponents by accusing them of dishonesty and hypocrisy, besmirch them publicly, and deflect attacks against him. A utilization of the whataboutism strategy occurred in an interview that Trump held on Fox News with Bill O'Reilly. At one point in the dialogue Trump equated American actions with those of Russian Vladimir Putin. O'Reilly challenged Trump by saying that "He's [Putin] a killer," to which Trump responded with the tactic of whataboutism: "There are a lot of killers. You think our country's so innocent?"[2]

This chapter zeroes in on the likely reasons why Trump's followers allow him to get away with this kind of deceit and verbal bellicosity. One of the main ones is that it is perceived as a courageous assault on political correctness and thus as part of the tactics required in the ongoing cultural war, which is being fought not with physical weapons, but with Machiavellian verbal weapons. As Rebecca Solnit has so perceptively argued, we may, in fact, be in the throes of a second Civil War in America, ignited by Trump himself who harkens back to a fantasy world of racial purity, with Trump as the first Confederate president since the war (born ironically in the Union state of New York).[3] The war is now being fought with verbal slings and arrows. But the danger is that words lead to actions, paralleling what happened in the Bolshevik Revolution, in European Fascism, and in Nazism.

BLUNT SPEECH

For preemptive strategies and counterattacks to be effective, they must be sudden, decisive, and "big" (recall the big-lie strategy discussed in the previous chapter). As the master liar knows, he cannot wait to react defensively to an attack on him; he must take care of any dangerous predicament preemptively and pounce on his attacker suddenly and brutally. It must be an "in-your-face" counterattack. This allows the liar to gain control of the adversarial situation. Machiavelli saw this as a performance of the lion role that the prince must always be ready to enact, in addition to being a fox who must detect an opponent's intent beforehand:[4]

> A prince, therefore, being compelled knowingly to adopt the beast, ought to choose the fox and the lion; because the lion cannot defend himself against snares and the fox cannot defend himself against wolves. Therefore, it is necessary to be a fox to discover the snares and a lion to terrify the wolves. Those who rely simply on the lion do not understand what they are about. Therefore a wise lord cannot, nor ought he to, keep faith when such observance may be turned against him, and when the reasons that caused him to pledge it exist no longer. If men were entirely good this precept would not hold, but because they are bad, and will not keep faith with you, you too are not bound to observe it with them. Nor will there ever be wanting to a prince legitimate reasons to excuse this non-observance.

Trump adopts the lion guise when he aggressively attacks the fake news media that critique him, pointing menacingly to reporters at the back of an auditorium during a rally, as "enemies of the people," or "evil people." This type of bellicose language has great resonance with acolytes and followers. It is a "sudden attack" language that contrasts noticeably with the polite discourse used by the reporters that he disparages. This is a key counteroffensive strategy, since it will be interpreted as a sign of strength rather than of boorishness. It is described by Machiavelli as follows:[5]

> If he [the prince] is wise he ought not to fear the reputation of being mean . . . and he can defend himself against all attacks, and is able to engage in enterprises without burdening his people; thus it comes to pass that he exercises . . . meanness towards those to whom he does not give, who are few.

Mussolini also adopted the lion persona, allowing him to counterattack his opponents by typecasting them as relativists who had the gall of believing that they alone were the bearers of "objective immortal truth," attacking fascism superciliously, which had the right to create its own ideology:[6]

> If relativism signifies contempt for fixed categories and those who claim to be the bearers of objective immortal truth, then there is nothing more relativistic than Fascist attitudes and activity. From the fact that all ideologies are of equal value, we Fascists conclude that we have the right to create our own ideology and to enforce it with all the energy of which we are capable.

This was a clever counterattack, making the "relativists" seem conceited and as enemies of freedom of thought. Mussolini also used blunt language that his followers found to be refreshing, in contrast to the sophisticated jargon of the relativists. Orly Kayam examined Trump's speeches, media interviews, and debates during the 2016 presidential primary campaign, finding a similar kind of bluntness, suggesting that Trump used this rhetorical strategy to gain popularity, tapping into the trend of anti-intellectualism that was fomenting in America.[7] Trump's blunt and direct language is not perceived as "uneducated" but as "honest," speaking to his followers directly, in contrast to the refined politically correct language of the "elite." British sociologist Basil Bernstein found that this type of style, which is called blunt here, emphasizes the "We" dimension of a social group, making members in it feel united, whereas an elaborated style, such as the one used by what Mussolini called the "relativists," puts the emphasis on the "I," and thus on the individual.[8] The former style fosters greater group adhesion; the latter does not.

Trump understood early on in the electoral campaign that for many "working people"—a term he used strategically over and over—politically correct speech was an evil antidemocratic trend, used by the elite to critique and demean "real" down-to-earth Americans. He pounced on every opportunity to drive this theme home in tweets, at rallies, and in front of the TV cameras. The profanities he uses are, thus, hardly perceived as coarse or uneducated, but rather as verbal weapons meant to be thrown into the faces of the elites. For instance, instead of using a statement such as "this is a deplorable situation," Trump would simply say or write "SAD"; instead of "this is nonsensical," he writes "STUPID!" This kind of in-your-face, blunt style is the opposite of the kind of measured speech that is normally expected of a statesman. Opponents decry his speech as "vulgar slang." But it is not slang in the normal sense of the word. As argued here, it is a military code that he uses antagonistically against his enemies, real or imagined.

The same type of speech was used to great effect by *Commedia dell'arte* actors to satirize pompous discourse styles. The *Commedia* was a popular Renaissance form of improvised street comedy based on plot outlines reflecting everyday life. It was intentionally humorous, vivid, crude, and often offensive. The actors understood that profane language had resonance with audiences, since it reflected the everyday reality of the people in their audiences, not the supercilious world of authorities and scholars. Like a *Commedia* character, Trump jumps out from the intellectual masses (or the relativists as Mussolini called them), speaking the "real language" of the people. His profane style of language, in fact, bespeaks of comedic theatricality—a pseudo-dramatic style intended to satirize the politically correct intellectuals, evoking laughter and implicit derision at the same time. When Trump finds himself in a formal context, such as when he is delivering a State of the Union speech, and where the use of blunt and profane style would be counterproductive, his delivery comes across as ineffectual and dry. The irreverent and coarse language that Trump uses at his rallies, on the other hand, has emotive force, never failing to evoke the laughter of derision from audience members. As the American writer Elizabeth Hardwick astutely observed, this type of speech "has the brutality of the city and an assertion of threatening power at hand. It is military, theatrical, and at its most coherent probably a lasting repudiation of empty courtesy and bureaucratic euphemism."[9]

Profanities are an intrinsic feature of the in-your-face approach that Trump utilizes so effectively. They have become so common in his tweets and rallies that they now hardly get noticed, and are largely ignored. When he referred to the Democratic House representative "Adam Schiff" as "Adam Schitt," it was barely critiqued by the pundits in the mainstream media, who have become indifferent to it:

> So funny to see little Adam Schitt (D-CA) talking about the fact that Acting Attorney General Matt Whitaker was not approved by the Senate, but not mentioning the fact that Bob Mueller (who is highly conflicted) was not approved by the Senate![10]

Perhaps in a mass media era, profane language may have lost its negative impacts. Glorified by movies and music videos, used throughout the internet, profane language affords people the opportunity to talk tough, just for the sake of it. In such use, however, the profanity becomes neutralized or at least diminished in its impact. The four-letter "F-word" is a case in point. It is used with regularity in media, in a matter-of-fact manner that hardly captures people's attention. Not too long ago, however, the word would have evoked negative reactions. The late controversial and brilliant comedian Lenny Bruce used it as a transgressive technique in his comedy act. Unlike many other comedians of his era, Bruce did not tell jokes. Instead, he attacked hypocritical attitudes toward sex, politics, and religion by speaking in a conversational manner, injecting frequent Yiddish words and profanities into his material, especially the F-word. Many were offended by his use of that word, and he was frequently arrested on obscenity charges. His speech clearly had a subversive impact; its use today in movies and television programs, on the other hand, has no such impact.

THE ATTACK ON POLITICAL CORRECTNESS

Profane speech is viewed by many of Trump's followers as an antiestablishment code and thus as more honest than the politically correct, hypocritical speech of the deep state establishment. Trump's fans admire his irreverent, earthy, barroom style as genuine and sincere. They find great delight in the subversive impact that his words have. Trump entered the debate on political correctness with a vengeance and, like the proverbial bull in the China shop, is seen as a destroyer of this insulting language. It is one of Trump's most effective lionesque strategies, since it is perceived as an important weapon in the insurgency against the repressive elite state, which promotes political correctness and inhibits free speech.

The attack on political correctness started with Allan Bloom's 1987 book, *The Closing of the American Mind*.[11] Bloom argued that the kind of speech that was being imposed throughout society, intended to avoid offending particular groups of people, was actually an ideological weapon of radical left politics, not a true antidote to racism, sexism, and social injustice. By claiming that it protected marginalized groups who were socially disadvantaged or constantly subjected to discrimination, it actually backfired since it stifled debate by not allowing opposite voices to be heard. It "closed the American mind," as Bloom decried. After Bloom's book, this type of "purity language" was called "politically correct speech."

There is little doubt that political correctness mobilized conservatives to attack it as a silly ploy of liberals that degenerated into an Orwellian system of mind control. As journalist Kat Chow has so perceptively observed, with the election of Donald Trump, the debate on political correctness became a fiercely emotional one, even though the term was forged in a different political context:[12]

> Since as far back as 1793, when the term appeared in a U.S. Supreme Court decision about the boundaries of federal jurisdiction, "politically correct" has had an array of definitions. It has been used to describe what is politically wise, and it has been employed as ironic self-mockery. The phrase has driven contentious debates in which free speech and free choice are pitted against civility and inclusion. But it hasn't just changed meaning, it has changed targets. What the November election [of Donald Trump] has made clear is that these words, especially when they're related to matters of multiculturalism and diversity, carry consequences.

The late president George H. W. Bush issued one of the first counterattacks on political correctness in May 1991, at a graduating class speech he delivered at the University of Michigan, aware that it had become a powerful counterstrategy in conservative ideology: "The notion of political correctness has ignited controversy across the land. And although the movement arises from the laudable desire to sweep away the debris of racism and sexism and hatred, it replaces old prejudice with new ones. It declares certain topics off-limits, certain expression off-limits, even certain gestures off-limits."[13] Bush's statement has much merit to it, and although the debate on how to be inclusive of otherness in a democracy has never been resolved, it is clear that language has always played a role in it. What Bloom's book did was articulate the fear of many conservatives that the liberal left in America had abandoned a basic principle of democracy—free speech. Political correctness was thus seen as indicating that a totalitarian nightmare was coming true, with the term *thought police* emerging as a common one in the 1990s to designate those who imposed their mode of politically correct speech on others, especially on campuses throughout the United States. As Dinesh D'Souza argued in his 1991 book, *Illiberal Education: The Politics of Race and Sex on Campus*, students in most classes had to conform to the supposed thought police's jargon that did not actually emphasize equality, but instilled a fear of being accused of promoting victimization and cultural appropriation unwittingly.[14] The book ignited a heated debate centered on identity politics, as it came to be known, as well as on the threat that PC (politically correct) language posed to freedom of speech. In a phrase, PC language was seen as a threat to American democracy.

The debate took on a life of its own, diverting attention away from substantive issues of otherness and diversity and focusing instead on words and their meanings—a debate from which, ironically, many linguists (including the present author) stayed away from, perhaps because the relation between language and politics has traditionally been a problematic one in the field. Nevertheless, a central principle of linguistics is that there is an intrinsic relation between language, culture, and thought. Changing words changes how we perceive things. As a classic example, consider a seemingly neutral word like *man.* In English, the word has meant, traditionally, "general human being." The problem is that the same word coincides with "the male person." The word actually meant "person" or "human being" in Old English and was equally applicable to both sexes. Old English had separate words to distinguish gender: *wer* meant "adult male" and *wif* meant "adult female." The composite forms *waepman* and *wifman* meant "adult male person" and "adult female person," respectively. In time, *wifman* evolved into the modern word *woman* and *wif* narrowed its meaning to modern *wife*. The word *man* then replaced *wer* and *waepman* as a specific term distinguishing an adult male from an adult female, but continued to be used in generalizations referring to human beings in general. The end result of merging these two meanings tended to render females invisible. The PC changes made to the English language over the last decades were attempts to correct this inbuilt semantic bias—*chairperson* (instead of *chairman*), *first-year student* (instead of *freshman*), and *humanity* (rather than *mankind*).

So, it should come as no surprise that political correctness emerged as a form of "linguistic activism" already in the 1970s, aiming to "correct" semantic structures in language that can potentially encode inequalities. Consider job designations as a case in point. Over the past seventy to eighty years, as women increasingly entered into traditionally male-based occupations, their presence was perceived (at first) to be a deviation from tradition. Logically, their job titles were marked linguistically by adding suffixes such as *-ess* to male-referencing words: for example, a female *waiter* was named *waitress*, and *actor* was named *actress*, and so on. It has taken decades to get across the point that the females in such jobs are worthy of the same words as those used for males (*waiter*, *actor*). And to this day, it is a battle to get such language changed to reflect the new realities. Moreover, changes in language do not always indicate the same kinds of social consequences. It all depends on the specific situation. Francophone feminists, for example, "advocate separate male and female terms because gender is an inherent feature of the French grammatical system."[15] Therefore, adding *-e* to the word *advocat* (male lawyer) to create *advocate* (female lawyer) is a linguistic validation of women's place in the professional workforce. However, because English grammar does not call for such gender agreements, we unnecessarily perceive terms such as *waitresses* and *actresses* as inseparable from their sexual identity.

As another case in point, consider the title *Ms*. It was introduced to correct a gender-based anomaly in the use of titles. The title *Mrs.* emphasizes a woman's traditional status as being married. *Miss* implies the opposite; namely, that the woman is not married. The term *Ms.* was introduced in the 1970s to rectify this double portrayal of women as either being "married" or "not married." The term provided a parallel designation to *Mr.*, which is not marked for marriage or lack thereof, thereby eliminating from the title the indication of a woman's marital status. However, as linguist Deborah Tannen points out, "Though *Ms.* declines to let on about marriage (*Mr.* declines nothing since nothing was asked), it also marks her as either liberated or rebellious, depending on the observer's attitudes and assumptions."[16] The solution might be, arguably, to address everyone with the same title, regardless of sex. That would indicate true equality.

The foregoing discussion was meant to illustrate why PC speech was initially a reparation mechanism to equalize social roles, including the recognition that women's traditional titles no longer properly represented their current realities. Paradoxically, titles such as *Mrs.* and *Miss* continue to be used today, in an era where gender equality in the workforce and in society generally is spreading. This indicates that old habits do indeed die hard.

Language is adaptive to social changes. This principle is almost completely absent in the debate on political correctness. Conservatives simply dismiss this PC language as catering to ideologically based identity politics, disregarding the equalizing mechanisms at work in language change; while liberals argue that the conservatives are using the debate simply to promote their own causes, leaving issues of racism, sexism, and social injustice unresolved. In a bizarre way, both sides make valid points, and this is why political correctness has evolved into such an emotionally charged debate—one into which Trump jumped to divide the two sides antagonistically, projecting himself to the front of the line as the leader of the forces against political correctness. All politically correct vocabulary is destroyed by Trump, with his bombastic attacks—"Bad," "Stupid," "Loser," and so on. This has attracted many conservatives to his side, seeing it as the perfect antidote to the PC mind control. It is one of his most effective strategies, since it indirectly declares war on the deep state, which is perceived as synonymous with "politically correct state." By leading the charge against this state, Trump is seen as a champion of freedom of speech. This became explicit when he made the following statement during a Fox News interview in 2016:

> I think the big problem this country has is being politically correct. . . . I've been challenged by so many people and I don't, frankly, have time for total political correctness. And to be honest with you, this country doesn't have time, either.[17]

As mentioned previously, at a rally in South Carolina, Trump called for "a total and complete shutdown of Muslims entering the United States"[18] to a cheering crowd, implying that this was probably not politically correct. This type of anti-PC stance projects him into the limelight as a fighter against the PC state and its purported Orwellian anti-free speech philosophy. His mission is to destroy it with his own profane, crude, and vulgar words and actions. He even used his anti-PC tactic during a primary campaign speech when he mocked a reporter with a physical disability. The crowd roared hysterically. He thus emerged from the primaries as a true fighter for "truth, justice, and the American way," as the tagline for the *Superman* TV program of the 1950s puts it.[19] He is the radical conservative movement's superman. His "flaws," like those of a superhero, are thus conveniently ignored, including the fact that he was a draft dodger, a no-no for conservatives in previous eras.

In a relevant 1992 essay, Ruth Perry warned that those who attacked people who used language that was meant to avoid offending others would eventually be able to manipulate social trends and lay the groundwork for a political countermovement.[20] The term *politically correct* was, in her estimation, recycled as a political weapon against people like herself who never used it to articulate injustices and biases in language. As Perry put it, "In the universities it is an attack on the theory and practice of affirmative action—a legacy of the sixties and seventies—defined as the recruitment to an institution of students and faculty who do not conform to what has always constituted the population of academic institutions: usually white, middle-class, straight, male."[21]

It is pertinent to note that in his book (mentioned above) Bloom attacked the Black Power Movement of the 1960s and 1970s as being actually counterproductive to the civil rights movement:[22]

> The Black Power movement that supplanted the older civil rights movement—leaving aside both its excesses and its very understandable emphasis on self-respect and refusal to beg for acceptance—had at its core the view that the Constitutional tradition was always corrupt and was constructed as a defense of slavery. Its demand was for black identity, not universal rights. The upshot of all this for the education of young Americans is that they know much less about American history and those who were held to be its heroes. This was one of the few things that they used to come to college [which] had something to do with their lives. Nothing has taken its place except a smattering of facts learned about other nations or cultures and a few social science formulas.

There is much to dispute in Bloom's statement, but it is beyond the scope of the present discussion to do so. Suffice it to say that the PC phenomenon has many sides to it (including the linguistic aspects mentioned above), and it cannot be reduced to simplistic disputations based on ideological agendas—to both the left and the right of the political spectrum.

Political theorists trace the modern-day notion of political correctness to Soviet communism in the 1930s, as a form of gaslighting. It was used to remind party members that the party's version of reality must be raised above reality itself. It was doublespeak in its most mind-controlling form. In the 1930s, Italian Marxist theorist Antonio Gramsci also saw it as an instrument of the state—ironically the fascist one in this case.[23] Referring to Machiavelli, Gramsci realized that human minds are not swayed by reasoning or by truth, but rather by contradictions and beliefs. These could be mobilized by the techniques of the master liar—be it an individual or the state itself (the Ministry of Truth). In chapter V of the *Prince*, Machiavelli observed perspicaciously that "the only secure way" to control people who have become accustomed to living in a certain way is to destroy that way, through subterfuge and mendacity and outright verbal attacks, eerily forecasting Soviet communism, fascism, nazism, and Trumpism.[24] Machiavelli's advice to the liar-prince was that he must use this strategy not to destroy people's minds, but to conquer them. In chapter VI he noted that there is nothing more difficult than to inculcate "new modes and orders," but if successful, the liar-prince will surely become "powerful, secure, honored and happy."[25] He clarified this insight in his *Discourses on Livy*: "When it happens that the founders of the new religion speak a different language, the destruction of the old religion is easily effected."[26] The Machiavellian prince, therefore, must speak a new language—a language that will persuade people to follow him. In the case of Trump, that language is the anti-PC one that he employs so effectively. Machiavelli saw language as the most powerful weapon for infiltrating minds and controlling behavior. To paraphrase Gramsci, the most effective "unarmed prophet" is the one capable of swaying minds with words.[27] Significantly, Gramsci was writing these words from jail, as his jailor, Mussolini, was forging a partnership with the Church, assuming the role of the unarmed prophet—in the same way that Trump is seen by evangelicals as their own prophet.

In sum, the attack on PC is a strategic one, with many historical parallels. By destroying the restrictions that conservatives saw as imposed on them by the PC police, Trump surfaced as a valiant warrior, who has allowed anyone to break taboos and mock whoever they want. He is perceived as restoring, through the example of his own blunt speech, the liberty to speak as one pleases. The fox has emerged as a lion, in the eyes of many.

ATTACKONYMS

As an anti–political correctness superhero, many see Trump as the only chance for America to be liberated from the PC police. Everything in his speech is antiestablishment, anticorrectness, and antinorms, raising him to the level of the mythical American rebel who resists authority, control, or convention. Even his use of misspellings is part of this brilliant strategy, at the same time that he uses them ingeniously to attack his opponents. At a surface level it appears that he is uncultured and even illiterate. But this is hardly all there is to it. The misspellings tap into an unconscious paradigm of rebellion against the establishment and its hypocritical norms and rules. Moreover, misspellings deliver subtle messages, as can be seen in the following tweet:

> Democrats can't find a Smocking Gun tying the Trump campaign to Russia after James Comey's testimony. No Smocking Gun . . . No Collusion." @FoxNews That's because there was NO COLLUSION. So now the Dems go to a simple private transaction, wrongly call it a campaign contribution.[28]

Needless to say, the correct spelling, as Trump would surely know, is "smoking gun." The fact that he repeated the same misspelling twice indicates that he knew what he was doing. The misspelled phrase mocks the "smoking gun" metaphor used repeatedly by the mainstream media in their reportage of the investigations against him. The phrase *smoking gun* implies the search for that one piece of incontrovertible evidence that would lead to his incrimination. So, the misspelling is satirical mockery and a counterattack strategy at once, since a *smock* is a loose garment worn over one's clothes to protect them—alluding to the superficial clothing of the investigations against him. It is a vulpine strategy that is based on suggestiveness and, like his other speech violations, is seen as part of the overall attack on the political correctness that is perceived as suffocating America.

Trump's use of hurtful nicknames is also part of the rebel-hero game that he knows how to play so proficiently. There is nothing more anti-PC than to call women "dogs" and "pigs" openly and bombastically. In the following tweet, he attacked his previous aide Omarosa Manigault Newman as a dog, receiving only scattered negative responses from pundits across media:

> When you give a crazed, crying lowlife a break, and give her a job at the White House, I guess it just didn't work out. Good work by General Kelly for quickly firing that dog![29]

His use of *dog* to attack women is truly offensive, since it alludes to a trope of women as sexual creatures who are intellectually inept. Yet his base perceives such speech as courageous anti-PC effrontery. Here are a few other examples:

> Robert Pattinson should not take back Kristen Stewart. She cheated on him like a dog & will do it again—just watch. He can do much better![30]

> Why is it necessary to comment on @ariannahuff looks? Because she is a dog who wrongfully comments on me.[31]

Using misogynist metaphors such as *dog* and *pig* would require a treatise on how these types of metaphors have arisen. Suffice it to say for the present purposes that they are part of an overall anti-PC attack strategy. During the first primary debate, interviewer Megyn Kelly confronted him with the following challenge: "You've called women you don't like fat pigs, dogs, slobs and disgusting animals." Trump interrupted her at that point, evoking audience laughter, saying sardonically but shrewdly, "Only Rosie O'Donnell."[32]

Trump knows that slurs and misogynist tropes are memorable and more likely to stand out because they are graphic and anti-PC, and they stick to the victim. As we may have ourselves experienced in school, if a smear or slanderous nickname was hurled at us, it would become something that we probably could never live down. School bullies use nicknames, smears, and slurs in the same way Trump does—to promote themselves by attacking others, thus deflecting any negative attention away from themselves. Insulting names are verbal attack weapons. They are among the most destructive of all tactics in the liar's arsenal.

A nickname is a means to pigeonhole someone, alluding to something in the person's character, appearance, or background that is thought to have significance (for better or worse). Trump's use of the adjective *little* in reference to Florida senator Marco Rubio ("Little Marco") and to Representative Adam Schiff, as discussed ("Little Adam Schitt"), not only refers to their height but also works at a different level.[33] Since they are "shorter" than Trump is, it also states, by implication, that they are endowed with "lesser" intellects than he is. At another level, the nickname alludes to someone who is relatively unimportant, having ironic intent. The ultimate objective is to evoke a negative image of a person's character and appearance.

The use of name-calling and nicknames is found in organized crime circles as well. Mafiosi have been long aware of the "brand value" of nicknames. Frank Costello, known as the "Prime Minister" of Cosa Nostra in the 1930s and 1940s in the United States, was quoted by *Time* magazine as stating this, as follows:[34]

> I'm like Coca-Cola. There are lots of drinks as good as Coca-Cola. Pepsi-Cola is a good drink. But Pepsi-Cola never got the advertising

> Coca-Cola got. I'm not Pepsi-Cola. I'm Coca-Cola because I got so much advertising.

Trump's nicknames are more than character profiles—they are character assassinations. One of his most widely used ones emerged during the presidential campaign, when he named his opponent, Hillary Clinton, "Crooked Hillary." This aims to remind his followers of her supposed illicit dealings in the past, which set off conspiracy theories from the alt-right that stuck to her throughout the campaign. By association, Trump is alluding to the entire Democratic Party, which she represented during the election and thus, by analogy, that the whole party was corrupt. In a penetrating article, Christian R. Hoffman compiled a corpus of around two thousand tweets to determine how Trump used this strategy in various situations.[35] The results showed that Trump was, as suggested here as well, adept at character assassination. Clinton's attack on Trump, on the other hand, fell flat, because she used the traditional verbal etiquette of American political discourse. The outcome of the election speaks volumes as to the effectiveness of Trump's verbal strategy. It certainly was not the only factor, but, as in a war, it was an effective one. For the sake of argument, this type of nickname can be called an "attackonym."

Trump's epithet for North Korean leader Kim Jong-un as "Rocket Man"[36] is an example of another kind of attackonym, whereby the name aims to belittle the person by portraying his character in terms of singular actions. Although it appeared to be a slur at first, it turned out to be a strategic ploy that got the "enemy" to the table (although the enemy in this case was more clever than Trump, ironically). At a literal level, this term made fun of Kim Jong-un's propensity to show off to the world by shooting rockets. But at a subliminal level, it arguably recalled Elton John's famous song "Rocket Man,"[37] which is about an astronaut lost in space, thus alluding that Kin Jong-un is someone who is lost. Actually, the same attackonym had appeared in reference to Kim Jong-un's father, in July 2006, when *The Economist* ran a cover story about him in

2006 with a picture of the dictator on the front cover showing him being launched from a rocket with the title, "Rocket Man."[38] It is truly ironic to note that Trump became, shortly after use of the attackonym, friendly to the North Korean dictator. The reason for this is unclear, but perhaps it is a strategy that fits in with his overall tactic of going against the previous political paradigm, showing himself to be able to interact with America's purported enemies better than any previous president.

Trump's reference to Republican candidate Jeb Bush during the primaries as "Low Energy Jeb" is yet another type of attackonym. It tags the former Florida governor as lacking the strength and vitality required for the sustained physical or mental activity of the American presidency. This was a death blow from which Jeb Bush never could recover. Its semantic potency lay in its low-key and thus subtle insinuation of someone who could hardly stand on his feet and thus part of a feeble Bush family when it came to toughness against enemies, from within or without.

Trump's use of attackonyms that begin with "lyin'" ("lyin' Hillary," "lyin' Ted," etc.) are both attack weapons and defensive strategies, since they also deflect attention away from his own blatant mendacity. How can a liar call someone else a liar? Sigmund Freud called this type of ploy *projection*—its aim is to lay on others what you yourself are.[39] The master Machiavellian liar is a master projector.

Trump has also used the "crazy" attackonym, a dog whistle, intended not only as an assault on the mental capacity of an opponent but also on the opponent's race, as is evident in his "crazy Maxine Waters"[40] nickname that he coined to attack the California Democratic Representative, who is an African American. His other attackonym for her, "low IQ Maxine," also constitutes a dog whistle, constituting a negative African American trope alluding to a purported lack of intelligence. He has similarly called MSNBC announcer Mika Brzezinski "crazed,"[41] and the adult film star Stormy Daniels, with whom he allegedly had an affair, "horseface,"[42] insulting not only her appearance but also brutally attacking her intelligence. In reference to Carly Fiorina, his only female rival during the primaries, he said the following at one of the debates: "Look at that face! Would anyone vote for that?"[43]

One final attackonym strategy can be mentioned here—creating a name that plays on the phonetics in a person's real name, through rhyme or alliteration. An example is Trump's nickname for former Arizona senator and former member of the House of Representatives Jeff Flake as "Jeff Flakey,"[44] in response to Flake's constant warnings about the dangers posed by Trump's presidency to democracy and its institutions. He called Democratic senator from Illinois Dick Durbin "Dicky Durbin,"[45] thus diminishing him through the suffix *-y*, which adds the nuance of "little" or "small" to a name. It also recalls the epithet given to Richard Nixon as "Tricky Dick," in reference to the dirty tricks Nixon used, culminating in the Watergate scandal.

As argued above, the reason why Trump gets away with this kind of slang is that it is perceived as having an anti-PC function. He has used the words *dumb* and *dummy* frequently to attack his critics, implying that he is the one who possesses true intelligence, especially with regard to women. He has attacked CNN announcer Don Lemon, an African American and one of Trump's staunchest critics, as "dumb" a number of times. This is a clear dog whistle referring to the myth that African Americans have a lower IQ. He reiterated this dog whistle in the following tweet:

> Lebron James was just interviewed by the dumbest man on television, Don Lemon. He made Lebron look smart, which isn't easy to do. I like Mike![46]

The fact that the tweet was retweeted over fifty thousand times immediately after it was sent speaks volumes as to the effectiveness of Trump's attackonym ploy. Trump's fans see it as both a defensive strategy and an attack on PC language, which would preclude ever attacking African Americans in such ludicrous and racist ways. So, it was seen as racist by some but as liberating by others—the division in American society could not be more marked. Lemon responded forcefully and aptly as follows:

> Referring to African Americans as dumb is one of the oldest canards of America's racist past and present: that black people are of inferior intelligence. This president constantly denigrates people of color and women.[47]

Attackonyms are, frankly, words meant to destroy others. It allows the liar-prince to appear as a lion, which, as Machiavelli knew, was an essential ploy in gaining and maintaining leadership. The prince can exert great power over his opponents, intimidating them with the possibility of using language against them. People know that once given a nickname, it is almost impossible to live it down. So, they tend to kowtow to the prince's whims in order to avoid being "name attacked" by him, fearing damage to their reputation.

Trump's attackonyms are overall part of his strategy of destroying the PC state. His slurs are seen as vindicarions for others to speak their minds. The irony is that most liberals have themselves attacked the PC affliction that started besetting America around the early 1980s. But it was the radical conservative media that portrayed it as a totalitarian tactic that was destroying free speech in America. Obama in particular was accused of being the worst perpetrator of PC talk and thus the undisputed leader of the PC state—a view that Trump instantly weaponized with his own style of blunt speech. As a master manipulator, he knew that the Zeitgeist was right for attacking PC, and thus that his blunt attacks on people would not be seen as misogyny or as defamation, but as bravery—the bravery to stand up to PC and demolish it with expletives. As Machiavelli suggested, a leader must resort not only to deception and disguise but also to bravado and feigned bravery to gain power and to protect himself against rivals. Trump came onto the scene when it was widely believed among conservatives that political correctness was eroding American liberty, inhibiting anyone to speak, act, and behave as they wished for fear of offending someone.

It is impossible to envision an effective counterattack to Trump's brutal verbal tactics given the anti-PC climate in which he devised it. The problem is that, while it is true that freedom of speech gives us the right to say what we feel, it also denigrates others. So, while Trump's fans see his blunt, profane language as liberating, there is a sinister side to it that might have brought about unwanted consequences, as CNN commentator Chris Cillizza so perceptively notes: "The problem with Trump's assault on political correctness is that he took it so far that he clearly emboldened not only those lurking in the shadows to bring their hate speech into the light of day, but also lowered the overall bar for what is considered acceptable discourse among politicians and other leaders in the country."[48]

DENIAL, DEFLECTION, DISTRACTION

Throughout chapter XVIII of *The Prince*, Machiavelli warns the prince that any admission of wrong-doing would be the end of his rule and control over people. So, to ensure that this does not happen he suggests the deployment of the strategies of denial, deflection, and distraction, which the prince should always have at hand to use against anyone who opposes him or presents evidence against him. These "3Ds" are effective counterattack and defense strategies. An example of the way Trump uses them is his shifting explanations about the so-called hush payments made by his former attorney Michael Cohen to sexual performer Stormy Daniels, with whom Trump supposedly had an affair long before he ran for president. First, he denied knowing about the payments. Subsequently, he deflected attention away from himself by answering a reporter's question about the payments, saying that she should ask Michael Cohen, who was his lawyer at the time. Finally, after Cohen testified against him, Trump mounted a series of attacks on Cohen, calling him a liar and thus deflecting attention away from himself while putting the focus on Cohen. At no time has he ever admitted to the affair for, if he did, and as Machiavelli warned, it would have been a disaster.

A few days before the November 2016 election, *The Wall Street Journal* reported that Karen McDougal, a former Playboy Playmate, had been paid off by the *National Enquirer*, who then did not publish the story so as to spare Trump the embarrassment that it would have entailed, and thus potentially affect public opinion against him during the election.[49] Even with mounting evidence and courthouse revelations of wrongdoing in this case, Trump denied everything, never once admitting that there was a romantic liaison between himself and McDougal. As a denial strategy, it was not enough to keep the media at bay, so after a while he simply started avoiding any questions from the media, distracting attention away from the scandal to other matters.

These two cases encapsulate how the 3D strategy unfolds—first deny something, when the denial falls apart, tell a new false version, while at the same time enacting distractions of various kinds, until the danger dissipates. Trump has applied the same three-pronged strategy to keep the public in a state of confusion with regard to his business and financial interests, as well as his alleged connections to Russia during the campaign. When these could no longer be denied, he simply asserted that there was no "big deal" to them, just something that any businessman would do. Slowly the story disappeared from the front pages, as Trump clearly anticipated.

As Jennifer Mercieca has perceptively pointed out, Trump's 3D strategy is part of an ancient rhetorical art that was called *apologia* by the Greeks—the art of verbal self-defense.[50] She observes that his defenders utilize another rhetorical strategy called "points of stasis," which are now called "talking points." These allow them to reframe a situation in such a way that it allows Trump to wiggle out of the difficulty. The strategy is to change the perspective, as Trump did by putting the spotlight on Cohen. As Mercieca aptly remarks, these talking points are designed to deflect attention away from Trump and to create obfuscation and doubt—there's no allegation of wrongdoing by the president; if he paid the hush money, it is not illegal; moreover, he cannot be indicted; and so on.[51]

Perhaps the most effective of all the 3D tactics in the *apologia* strategy is deflection, also known as "whataboutism." The objective is attack or discredit critics by charging them, or others, of the same thing of which they are accusing you, without refuting the initial attack or disproving it in any way. This turns defense into offense: "What about X?" is the most common of all deflection strategies used by both Trump and his surrogates. Although the origin of modern-day whataboutism is traced to Soviet Russia, it has its roots in Machiavelli, who saw this particular aspect of *apologia* as a particularly important one, because his supporters are willing to be deceived (already cited but repeated here for convenience):[52]

> It is necessary to know how to conceal this characteristic well, and to be a great pretender and dissembler. Men are so simple, and so subject to prone to be won over by necessities, that a deceiver will always find someone who is willing to be deceived.

The Machiavellian prince is a master of deflection, able to make spurious claims in order to deflect the spotlight away from his demonstrable falsehoods, turning the spotlight on his opponents. When Trump is at a loss of words in the face of a counterattack, he has developed a repertoire of deflective stock phrases that allow him to evade direct criticism, postponing responses into the future, where they will never be answered. These include:

> "A lot of people are saying . . . "
> "Everyone knows . . . "
> "A lot of people tell me . . . "
> "I've heard that . . . "
> "People think it's going to happen . . . "
> "Everyone is now saying . . . "

His vacuous claims of voter fraud, his suggestion that Obama wiretapped Trump Tower during the election, but offering no evidence, are examples of how he uses deflection to protect himself from accusations, redirecting news coverage to his claims of harassment or the perpetration of hoaxes against him. This is pure Orwellian doublespeak, designed to create doubts and confusion and thus provide Trump with verbal shields. As with other forms of doublespeak, it disrupts peace of mind, while shielding the liar con artist. Ken Kesey's 1962 novel *One Flew over the Cuckoo's Nest*[53] takes place in a psychiatric hospital where a con man rules the asylum. It is essentially a cautionary tale—we are all susceptible psychologically to the manipulations of con men, who may have imprisoned us in our own mental asylums. A con man like Trump inevitably disorients his enemies, catapulting them into a cuckoo's nest of his making. As the saying goes, "Words matter." They cannot be ignored because of their ability to spur people on to action and to disrupt peace of mind. It is naïve to say that what counts is policy, as Trump defender's claim. Words affect the sanity of mind.

WORDS MATTER

An aside to emphasize the importance of the last point might be useful, since we may be unaware of how doublespeak can affect the operations of the mind. There is psychological research that indicates that words influence the diagnostic and healing processes. In one relevant study, Casarett, Pickard, Fishman, Alexander, Arnold, Pollak, and Tulsky found that physicians use metaphorical language in diagnostic interactions around two-thirds of the time. When they did, patients reported that they were able to grasp the implications of the disease and its symptomatology much more tangibly.[54] Some metaphors are, of course, better than others. The "cancer is war" metaphor seems to have therapeutic resonance in Western societies, given its diffusion in everyday language.[55]

In a truly fascinating book on the use of metaphorical discourse in clinical therapeutic situations, Linda Rogers provides strong evidence that the function of such discourse is, in effect, to promote understanding on the part of the patient.[56] In therapy sessions with one particularly troubled and disabled patient, JR, Rogers used the "life is a journey" conceptual metaphor ("You have a long way to go," "Your life is ahead of you," "Don't look back," etc.) to help him learn how to cope with his condition in a socially acceptable fashion. JR came to Rogers because he had sought work over a three-year period that was consistent with his real abilities, but he could never get the job he wanted. Rogers discovered that his lack of success was not due to any lack of skill, but rather, to the fact that he could not negotiate social discourse successfully. The journey metaphor allowed JR to understand what the underlying problem was through analogical reasoning. Rogers was thus able to help JR out of his dilemma by bringing him to the realization that, in expressing himself in discourse, he had to abide by specific rules of human social interaction.

In another case, Rogers used a similar technique to help a patient named Sarah overcome a deep trauma that resulted from being shot at by a deeply disturbed man while she was waiting for her husband to pick her up outside a university building. Rogers was in the building at the time teaching a class and, therefore, was able to run to her rescue, saving Sarah by applying artificial respiration and by talking to her. As her patient subsequently, Rogers helped Sarah recover from her emotional wounds by getting her to narrate the incident over and over—an incident from which she wanted to dissociate herself completely, as if it never happened. In her narratives, Sarah, a special educator, saw herself as wanting to help her assailant, who was later shot dead by the police during a confrontation. Sarah's discourse with her assailant revolved around how she would have helped him understand himself, and thus come to grips with his emotional confusion. But in her dialogue, the assailant was always a silent partner. Thus, her narrative was hardly cathartic. As a result, Sarah sought another kind of solution—changing her persona by changing her body image through tattooing. Using the same journey conceptual metaphor that Rogers used with JR, she was able to help Sarah come to the view that she could only come to grips with the situation by changing her path and starting a new journey. As in the case of JR, the strategy worked.

The objective of the foregoing discussion is to argue that words matter; they can upset or restore mental health.[57] It is no exaggeration to suggest that Trumpian doublespeak is upsetting the mental stability of many in American society. The use of the term *verbal weaponry* in this chapter to describe how he and others have used words to attack people, to offend them, as well as to shield themselves through doublespeak, has the intent of suggesting that, like weapons, words can harm, wound, injure, and damage people's sanity. By creating a "cuckoo's nest," he can take over the asylum much more easily than otherwise.

The American linguist George Lakoff has always stressed the power of metaphors to affect human lives: "Metaphors can be made real in less obvious ways as well, in physical symptoms, social institutions, social practices, laws, and even foreign policy and forms of discourse and of history."[58] The likely reason why metaphor is so effective cognitively, behaviorally, and emotionally is its source in what neuroscientists call "mental blends."[59] A blend is formed when the brain identifies distinct entities in different neural regions as the same entity in another region. Together the entities constitute the blend. In the metaphor of "fighting a war on cancer," the two distinct entities are "cancer" and "fighting." The blending process is guided by the inference that disease is a war, constituting the final touch to the blend—a touch that keeps the two entities distinct in different neural regions, while identifying them simultaneously as a single entity in the third region.

In her classic book *Illness as Metaphor*, the late writer Susan Sontag argued persuasively that people suffer more from conceptualizing metaphorically about their disease than from the disease itself.[60] There is an important warning in Sontag's assessment. Words affect everything, from politics and science to mental and even physical health. One should never be naïve about the effects that attackonyms, profanities, and doublespeak have on people's minds. The PC movement was, in a sense, a mental health movement, aiming to protect people from being demeaned. It may have been way too aggressive in its social-therapeutic goal, but it certainly did not upset the sanity of society. Trumpian doublespeak certainly does.

EPILOGUE

In a fascinating book, *Trump Revealed: An American Journey of Ambition, Ego, Money, and Power*, Michael Kranish provides an extensive list of examples of how Trump has used attackonyms and the *apologia* (3D) strategies opportunistically.[61] The most revealing example refers to an event before Trump entered the campaign when he was sued by the Justice Department for racial discrimination for refusing to rent his properties to African Americans. His counterattack was forceful—he accused the Justice Department of defamation, turning the Department's accusation of racism against him to one of discrimination against him, Trump. As a result, the Department ended up settling the case out of court. Trump learned how language can be used in a hostile way to get what he wants or, at the very least, to shield himself. As Machiavelli certainly understood, verbal weaponry is critical for self-preservation, perhaps more so than military combat.

The verbal weapons discussed in this chapter constitute a relatively small selection of those that the liar-prince can use to great effect, both to attack and counterattack opponents. The fact that they can be described concretely and that we can see through them perspicaciously raises the question: Why are so many people duped by them? As mentioned several times already, the plausible reason is that they are perceived as weapons in the greater war against the deep state—weapons that Trump needs to defend himself against his own attackers. This was, to reiterate, an overriding theme in Machiavelli's manifesto—gaining strength with language that is both aggressive and defensive is one of the most important of all skills in the Art of the Lie.

A study by psychologist Thomas Pettigrew sheds light on this Machiavellian strategy.[62] His overall findings can be summarized with the proverb "All is fair in love and war." Pettigrew calls the type of mind manipulation that Trump is able to perpetrate "Social Dominance Orientation," which refers to the knack of some people to see authoritarian figures as necessarily born to be leaders. Pettigrew found that those who favored Trump in the election saw him as a born leader, ready to bring down the PC state. The cathartic relief expressed through laughter at Trump's rallies is indirect proof that he succeeded in assuring his followers that he would indeed set things right in the world. During a 2015 interview on the CNN news program *Meet the Press,* it became saliently obvious that Trump would use his type of language as his main weapon in his assault on the PC state: "We have to straighten out our country, we have to make our country great again, and we need energy and enthusiasm. And this political correctness is just absolutely killing us as a country. You can't say anything. Anything you say today, they'll find a reason why it's not good."[63]

An interesting and highly relevant 1943 short film titled *Don't Be a Sucker* was produced by the United States Department of War and the Warner film studio.[64] It was intended to warn against the rise of Nazi sympathizers in the United States, cautioning people to beware of the language used to entice them into their cuckoo's nest. A young Freemason is shown being duped by the persuasive language of a soap box orator who claims that all the "good jobs" in America were being taken by minorities, domestic and foreign. The young man then initiates a discussion with a refugee professor who warns him that the same pattern of talk brought Hitler to power in Germany, splitting the country into groups, each hating the other. As a cautionary tale, this film reverberates with great significance in the current climate. Its moral is self-evident.

As Machiavelli emphasized, the main ploy is to make sure that "few really know what you are."[65] The liar-prince must always appear to be a lion but think and act like a fox, so that those who see through him are rendered immobilized through such strategies as the 3D one. Machiavelli's hero was Cesare Borgia (c. 1475–1507), the political leader who espoused and utilized the Art of the Lie as a Machiavellian fox, at the same time that he acted upon the social stage as a ruthless lion. Machiavelli praised Borgia in *The Prince* as the model of an unscrupulous leader who knows how to keep power. One time, when Borgia's enemies started to plot against him, he captured them by feigning to be friendly to them and then had them murdered. He also had his sister's husband assassinated for political reasons. Machiavelli's ideas were generalized by British philosopher Thomas Hobbes, who characterized human life as a "ceaseless search for power."[66] It is obvious that Trump is a perfect example of someone who is in a ceaseless search for power, using any means available to him to attain it—no matter how many casualties are involved.

Most people, observed Machiavelli, do not look to causes, but to effects. This is why Trump constantly claims that he has kept his promises to reform American society and restore its past grandeur. To end this chapter with a summative cliché, it can be said that if Trump had put a gun to people's heads, he would not be around; but by putting words to their heads instead, he rose to power with the support of those beguiled by his words.

7
TRUTHFUL HYPERBOLE

Without promotion, something terrible happens... nothing!
—P. T. Barnum (1810–1891)

PROLOGUE

As mentioned in the preface, Trump's book, *The Art of the Deal*, contains the phrase *truthful hyperbole*, which is defined as language that plays "to people's fantasies" and allows those who "may not always think big themselves" to do so.[1] The strategy reflects a worldview that can be encapsulated as follows: "to think big means to talk big."

This philosophy does not originate with Trump; it starts in the nineteenth-century American business world of "wheeling and dealing" that adopted bombastic language to promote goods and services, imitating the loud "big talk" of circus impresarios, of spectacle announcers, and of other showmen who would literally shout their messages and appeals to their audiences. Big talk is persuasive and coercive—constituting a form of verbal arm-twisting. The historical figure who may have introduced such talk into society as a form of common discourse, rather than as exceptional oratory, was the legendary entrepreneur, showman, and circus operator P. T. Barnum (1810–1891), who understood how big talk could stir up emotions and persuade people to do things that they otherwise would not do. As a result of his many successes, Barnum's hyperbolic speech became a dominant style in the marketplace and in advertising, enticing people to come forth and enjoy the delights of shopping and of spectacles like the circus and vaudeville.

Barnum introduced such stock hyperbolic phrases into the American business lexicon as "Don't miss this once-in-a-lifetime opportunity!" and "An unbelievably low price!" At the "Greatest Show on Earth," as Barnum strategically called his circus, people could find something to amuse themselves—it was a promise, not a fact. Barnum's influence on the language of America cannot be underestimated. Hucksters, con artists, and cagey politicians have realized that hyperbolic speech allows them to gain people's trust in the same way that Barnum was able to gain their attention. Hucksters use big talk to sell anything to anyone, and especially to promote themselves as larger than life. Trump is a descendant of the bombastic huckster figure, and his *Art of the Deal* is, in part, a modern-day manifesto of hucksterism. He knows that bombastic, hyperbolic language stirs people's emotions. So too did Mussolini and Hitler, albeit in different contexts. At rallies, Mussolini ignited cries of support with his loud grandiloquence, prodding the audience to respond hysterically with "Duce, Duce, Duce!" Hitler also aroused the emotions of his audiences in a similar way, impelling them to shout "Sieg Heil, Sieg Heil!" At Trump's rallies the "Lock Her Up!" shouts of the audience fall into the same category of hyperbolic elicitation.

This chapter looks at the language of hyperbole, which Trump uses to great effect to sell his ideas, falsehoods, and "bullshit," a profane term that I cannot avoid in this chapter, even though I find it myself to be distasteful. The problem is that truthful hyperbole has not stopped at the boundaries of Trump's rallies. It has migrated across the social spectrum, and in particular to cyberspace. The promise of "expressive freedom" that the Web 2.0 world was supposed to have made possible has turned into a world of "truthful hyperbole" and "alternative facts" that go largely unnoticed for what they are—gimmicks of self-promotion. Hucksterism, bombast, and bullshitting have spread surreptitiously across all digital platforms, replacing genuine conversation. The objectives of this type of language include scams, flimflams, rip offs, hoaxes, conspiracies, and so on. In the world of the matrix, truthful hyperbole exists side by side with genuine speech, and the difference between the two is becoming less and less relevant.

Trump's big talk is not seen in a negative light by his fans, but rather as a style of discourse that is quintessentially American, and a weapon against the PC establishment (as argued in the previous chapter). His purported success as a businessman also fits in with this American mythology—he is perceived as shrewd and relentless in his own realization of the American Dream. In Trump country, he is "one of them," a down-to-earth dealmaker who, by his actions, will make economic and social life better for everyone. Anyone who does not subscribe to this worldview is seen as an outsider, and thus as "un-American," as Trump has so cleverly suggested in his many rallies and tweets. He has come to town to run the "greatest show on earth," where hope and promise are bandied about speciously without any proof that they will ever become realized.

As Claudia Claridge has cogently argued in her book *Hyperbole in English: A Corpus-Based Study of Exaggeration*, hyperbolic speech is effective today because it is everywhere.[2] Using data from everyday conversations, TV programs, newspapers, and literary works, Claridge concluded that hyperbole is hardly an exception to discourse, but rather a powerful form of persuasive language that works on people below the threshold of reflection. As in some business deal pitches, it inflates, exaggerates, overstates—all in the service of selling something to someone by magnification of content. The bragger is thus seen as having swagger, not arrogance.

Actually, hyperbolic speech has always been part of the traditions of oratory, which is said to have been founded in the city of Syracuse in Sicily in 466 BCE, by Corax, a Sicilian Greek, who established a system of rules for public speaking with the help of his pupil Tisias. During the 400s BCE, Athenian citizens attended the general assembly, where public policies were announced and debated, participating directly in the administration of justice. In the courts, a verdict was argued by debaters, and each case rested with the jury, since there were no judges. This led to a focus on oratory and what made some words persuasive. Followed by the works of the Roman Cicero, the first renowned orator, the idea was to understand why oratory was a form of persuasive argument. It can be argued that without hyperbolic speech, dictators and autocrats would hardly ever gain the upper hand. The rise to power of Mussolini, Hitler, and Trump is evidence in support of this principle of oratory.

A LANGUAGE OF BUSINESS AND RELIGION

Barnum's influence on the social evolution of America cannot be understated. His style of hyped-up speech, part business sales pitch and part religious revival fervor, is used in advertising, marketing, public relations, media, and certainly on the internet to promote anything and everything. Slogans and taglines are forged with this very style. The following observation by Marty Neumeier with regard to the Nike slogan, "Just do it," brings out perfectly why such language is effective:[3]

> As a weekend athlete, my two nagging doubts are that I might be congenitally lazy, and that I might have little actual ability. I'm not really worried about my shoes. But when the Nike folks say, "Just do it," they're peering into my soul. I begin to feel that, if they understand me that well, their shoes are probably good. I'm then willing to join the tribe of Nike.

As Neumeier suggests, truthful hyperbole is effective because it brings with it a promise, and thus seems to talk directly to the prospective buyer, rather than presenting information about a product in an abstract descriptive way. This type of language when used in common everyday speech has conditioned all of us to think that we are important and we too can become part of "big thinking," through hyperbolic slogans such as "so much," "great," "very," and "tremendous," which are common superlatives used by Trump and previous Machiavellian masters of the lie. The following statements were made by Trump during the primaries and the election campaign:

> "We will have so much winning if I get elected that you may get bored with winning."[4]

> "I will build a great wall—and nobody builds walls better than me, believe me. I will build a great, great wall on our southern border. Mark my words."[5]

> "I think I did a great job and a great service, not only for the country but even for the president in getting him to produce his birth certificate" (spoken in reference to his bogus birther claim).[6]

> It would . . . "create tremendous numbers of new jobs" (in reference to his tax cut).[7]

In this kind of hyperbolic speech there is a constant subtext: "Things are *bad* and they will be made *better* and even *great* by me, Donald Trump." Barnum was a master at this kind of hyperbole, using it for promoting the "good life," melding business *esprit* with religious zeal. Barnum's hyperbolic language was akin to revivalist religious oratory. Barnum's goal was money making, but he also saw this as part of a religious "ministry":[8]

> This is a trading world and men, women and children, who cannot live on gravity alone, need something to satisfy their gayer, lighter moods and hours, and he who ministers to this want is in a business established by the Author of our nature. If he worthily fulfills his mission and amuses without corrupting, he need never feel that he has lived in vain.

This blurring of the line between business and religion is also common at Trump's rallies, which revolve implicitly around moral decline and racial purity, as S. Romi Mukherjee has perceptively pointed out: "Trumpism's narrative of decline is bound to the narrative of white and White-Christian decline, reconfiguring 'Americanness' in terms of imaginary racial purity."[9] Like Mussolini's and Hitler's charismatic speeches at rallies, Trump's oratorical splurges are akin to a preacher shouting out his promise to eliminate "sins" (diversity, racial equality, sexual diversity, etc.) at a mass revival, inciting hysterical outpourings in the congregation, with his promise to save the nation and make it great again. A similar observation was made by Michael Wolff in his book *Fire and Fury*, in which he compared Trump's rallies to "big tent revivalism," thus alluding to the blend of the big talk of circuses and religious revivals.[10] Trump's "big talk" is thus felt to be spiritually cathartic, with its promise to overturn the PC state and to retrieve America's true past.

In America, the businessman and the preacher have typically used the same kind of speech tactics. The Trump rally is, in effect, a "big tent revival," a spectacle that enmeshes folksy-style preaching proclaiming the good news with hyperbolic attacks on the fake news and the enemies within America. It is revival spectacle that is intended to evoke applause, laughter, and religious fervor. Trump's hyperbolic oratory has allowed him to sculpt himself as a "White Jesus," who promises prosperity and spiritual redemption at once, as Chauncey DeVega has perceptively pointed out:[11]

> Trump has become a "White Jesus," a pseudo-Christian savior, to whom evangelicals offer their votes, their allegiance and their political donations, worshiping at the altar of this false image of God. American conservatism at present is deeply fundamentalist. But it is also deceptively inclusive: authoritarians, bigots, racists, misogynists, white supremacists, nativists, gangster capitalists, the willfully ignorant and anti-intellectual, and those who eschew reason for passion are all welcome.

Trump's followers extend beyond the realm of evangelicals, of course, but the gist of DeVega's assessment is valid. Trump's motivational rallies are similar in delivery style and tone to those of preachers like the late Billy Graham and business guru Tony Robbins wrapped into one. They speak not to reason and objective truth, but to the emotions, aiming to liberate pent-up feelings of fear and resentment that many had suppressed before Trump came to office. His rallies can be characterized as "orgies of feeling," an appropriate term used by Elisabeth Anker.[12] It is no coincidence that Pentacostalist preacher Paula White claimed that Trump's presidency was anointed by God on the *Jim Bakker Show*, in order to bring America back to its Christian roots.[13] This is arguably why, as Michael Wolff has also suggested, that a Trump rally is hardly just a political event, but is perceived as a clarion call to the restoration of White religion and a cathartic outlet for White anger. Trump is thus seen as the charismatic traveling preacher maverick coming to town to deliver the good news.

As Reza Aslan has written: "Trump has harnessed the kind of emotional intensity from his base that is more typical of a religious revival meeting than a political rally, complete with ritualized communal chants."[14] Interestingly, Barnum was also perceived as a motivational, revivalist speaker, who actually was a fanatical supporter of temperance. Barnum was a believer; there is no evidence to support that Trump is as well—he is a "Dottore" character (as discussed below). Nonetheless, the parallels between Trump and Barnum are striking. In addition to his activities as a showman, Barnum became active in politics—prefiguring the contemporary blurring of the lines between statesmanship and showmanship. He was elected to the Connecticut legislature in 1865 and 1866, and served one term as mayor of Bridgeport, Connecticut, in 1875 and 1876. He was also a tenacious and dogged pro-Prohibition lecturer and author, writing *The Life of P. T. Barnum*,[15] one of the most popular autobiographies in American history, where the persona of the huckster is sculpted in the words of one of America's greatest hucksters of all time.

In a discerning book, *Bunk: The Rise of Hoaxes, Humbug, Plagiarists, Phonies, Post-Facts, and Fake News*, Kevin Young also sees striking parallels between Barnum and Trump, since both were masters at perpetrating hoaxes and fake news to great personal advantage, while blaming others for these exact same strategies.[16] Barnum, like Trump, was keenly aware of the power of the press to influence the opinions of the masses and thus made sure that he could outdo the sensationalistic language that the yellow press adopted at the time, eclipsing that style in his own ways. Like Trump, Barnum also planted fake stories in the press every once in a while to impugn the validity of the press itself. Both experienced bankruptcies and still ran successfully for office. Barnum used the circus as his platform for promoting himself politically; Trump used Reality TV. As Young points out: "As viewers, we inheritors to Barnum's America tend to feel a mix of I can't believe I'm watching this, and I can't believe that person did that, to I can't wait to see what happens next."

Barnum ironically decried the state of moral decay in America at the same time that he extolled its excesses—an approach that Trump has adopted as well. In the presidential debates, he continually blamed the rise of ghettoized inner city communities on corrupt politicians (the Democrats) and that only he can solve their problems. Self-promotion is an essential part of the sales pitch, as Barnum also knew. It needs no justification; by simply claiming it to be so it sticks to people's minds.

As mentioned briefly in chapter 1, James Pennebaker was able to empirically establish a link between pronouns and personality.[17] Pennebaker examined the speeches of several American presidents (previous to Trump) and found, for example, that Barack Obama was the lowest first-person pronoun user of any of the modern presidents—that is, the personal pronoun "I." Pennebaker argued that when presidents used this pronoun abundantly in their speeches their intent was to personalize their message, conveying to audiences that they were committed personally to specific causes. Obama's apparent avoidance of this pronoun did not mean, however, that he was humble, insecure, or uncommitted; on the contrary, it showed confidence, self-assurance, and a high degree of commitment in an implicit and suggestive way. The gist of Pennebaker's study is that a simple pronoun reveals more about personality than do content words (nouns, adjectives, verbs) or any psychological profile. Pronouns have an "under-the-radar" meaning to them, constituting traces to what is going on in someone's mind. In striking contrast, Trump uses the "I" constantly, projecting himself into the limelight, implying that he, and he alone, can be the Pied Piper leading people out of the very mess that he attributes to Obama. As an aside, my use of the Pied Piper analogy is intentional, since the ending of the story is a premonition. "The Pied Piper of Hamelin," as is well known, was an 1842 poem by Robert Browning, based on an old German legend. The piper, who was dressed in colorful costume, rid the town of Hamelin of its rats by enticing them away with his music, and when refused the promised payment he lured away the children of the citizens to their doom.

It is interesting to note that Pennebaker has looked at diary entries written by subjects suffering through traumas and depressions of various kinds. He discovered that pronouns were indicators of mental health, claiming that recovery from a trauma or a depression entailed a form of "perspective switching" that pronouns facilitated. He also found that younger people and those from lower classes used "I" more frequently than others, indicating a socially and psychologically constrained perception of Selfhood. His work has received significant attention both on the part of linguists and medical practitioners, who use what Pennebaker calls "expressive writing" for therapeutic reasons. This line of research puts the spotlight directly on Trump's constant use of "I" as a strategy of differentiating himself from previous presidents, but at the same time it suggests the kind of mental instability that is symptomatic of narcissism (as will be discussed below).

It is worthwhile taking a closer look at the strategies behind Trump's hyperbolic ploys. During an election debate he referred to violent crime in Chicago as follows: "In Chicago, they've had thousands of shootings, thousands, since January first. Thousands of shootings."[18] As an indeterminate number, "thousands" achieves several semantic objectives at once—it paints an image of crime as rampant; it evokes scenes of crowds of criminals or gang members (a subtle dog whistle pointing to African Americans); it paints a picture of a society in chaos and thus, in need of a strong leader who will impose "law and order" and eliminate the "lawlessness." A common ending to his tweets and various speeches, as an emotional punctuation point, is "Bad," "So bad," or "Really bad." Again in reference to crime, he said during an election debate: "Here you have so many bad things happening, this is like medieval times"[19] The number of unconscious connotations that this hyperbole evokes suggestively is infinite. If something is bad, such as the state of affairs engendered by the deep state, then members of that state are the ones who have left a "mess," as he has often put it or else a "disaster," as he excoriated during a presidential debate: "We invested in a solar company, our country. That was a disaster. They lost plenty of money on that."[20] A key feature of his linguistic exacerbations is that they are not isolated, but form a kind of rhetorical lexicon. Like a town crier or preacher, Trump thus establishes a chain of controlled connotations through a thesaurus of hyperboles, which lead followers to accept his conclusion that "winning" can only be achieved through the great deal maker: "I know how to win."[21]

Research in linguistics has shown that the ways in which people talk not only taps into a system of implicit social meanings and connotations but also shapes and changes the interpretation of words themselves.[22] Trump's systematic and repeated use of hyperbole is a powerful form of mind-shaping speech that affects the emotions and belief in the same style of revivalist religious discourse. The businessman-preacher is someone that many see as an American hero. Some of those in his base may see through his bombastic verbiage, but they accept him nonetheless as their leader because he is delivering the highly conservative agenda that they desire. Trump shrewdly presents himself as the only one able to carry this out.

NARCISSISM

To some, Trump's constant, bombastic promotion of himself is evidence that he may be afflicted by a disturbing form of narcissism. The anecdotal evidence for this comes primarily from his rambling tweets and rally rants. So, rather than the language of the Barnum-style businessman-preacher, Trump's verbiage may be symptomatic of a narcissistic personality disorder, which consumes the individual to the point that he feels the need to eliminate anyone who stands in his way. American psychologist Theodore Millon identified five subtypes of narcissism that induces someone to develop a misplaced and unfounded sense of grandiosity and entitlement.[23] Trump's behavior and language can be seen to manifest the symptoms of all the subtypes, which are listed below:[24]

The Unprincipled Narcissist: The main features associated with this type of narcissism, all of which are manifest in Trump's words and actions, include disloyalty to others, erraticism, arrogance, and vindictive behavior. Psychologists have found that an unprincipled narcissist is typically a swindler, an embezzler, and an unscrupulous person. Trump's vindictiveness against those who have testified against him, such as Michael Cohen, is truly worrisome, as is his arrogance and disloyalty to anyone who confronts him or does not do his bidding, including such ex–White House workers as James Mattis and Omarosa Manigault Newman, among many others whom he has dismissed from the government.

The Amorous Narcissist: The characteristics associated with this type of narcissist include the fact that he is a superficial charmer, a glib smooth talker, and a hedonistic indulger whose own pleasure and need for carnal satisfaction always come first, even at the expense of others. Needless to say, Trump's treatment of women is a classic textbook portrait of the amorous narcissist. As one of his paramours before he became president, Karen McDougal, stated during an interview on

CNN with Anderson Cooper, Trump seduced her with his charm and smooth talk.[25] It became obvious to her over time, however, that he was only interested in his own pleasure and sense of romantic conquest, rather than being truly "amorous."

The Compensatory Narcissist: The main characteristics of this type of narcissist is his paranoia over his own self-esteem. He is plagued by feelings of insecurity and thus will feign expertise and project a false bravado or veneer of superiority to shield himself against criticism or failure. He is typically overbearing, a micromanager, and a charlatan. Trump is constantly feigning expertise, claiming that he knows "more than the generals" or the economists that point out the flaws in his approaches to the economy. His constant overturning of the White House staff clearly indicates that he is an obsessive micromanager.

The Elitist Narcissist: The elitist narcissist is exploitative of others, generally hailing from a privileged background, having a sense of entitlement by birthright. This type of narcissist is enabled by parents during the formative years. As reported by *The New York Times*, Trump is hardly a self-made man.[26] Trump boasted during the election that he started with virtually nothing, hiding the fact that his father helped him financially from the beginning.

The Malignant Narcissist: The malignant narcissist is aggressive, angry, vengeful, cruel, seeing the lives of others as inferior or trivial. There is little need to elaborate on this trait here—literally all of Trump's presidency is proof of his malignant narcissism.

Another subtype to the above typology can be added—namely, in a world of selfies, Facebook profiles, and the like, there is a *Silent Narcissism* that has become virtually normal in cyberspace. Everyone can be a little larger than life through truthful hyperbole on social media. Cyberspace has had a powerful role to play in allowing Trump to get away with his narcissism because it has become "silent"; that is, narcissism now goes largely unnoticed or at least ignored, having evolved into an unconscious character trait in the world of the matrix. In the pre-Web 2.0 era, Trump's narcissism would have been considered boorish at best and insane at worst. The silent narcissism of the Web makes any attack on his narcissism ineffectual—it is considered simply to be part of his persona, not a clinical state of mind. We are living in a golden age of narcissism, and Trump simply blends into it.

The lexicon of the narcissist is no longer seen as braggadocio, but part of big talk—"tremendous," "bad," "mess," "thousands," "disaster," "unbelievable," "terrible," "winning," and so on. During the presidential debates, Trump referred to his ability to make money as his "tremendous income" ability, thus implying that he could apply the same ability to run the economy and thus "to bring tremendous amounts of money, tremendous amounts of jobs, tremendous numbers of companies."[27] In any previous era this would have been seen as the vacuous, boastful words of a narcissist, not a believable businessman. Today it is largely ignored. As Joseph Burgo has aptly put it: "To describe Donald Trump as a narcissist has become cliché, so widely accepted that the use of the label barely raises an eyebrow."[28] Burgo goes on to make the following insightful comment:

> The rise of Donald Trump thus marks the fusion of populism and narcissism. In times of enormous demographic shift and economic uncertainty, populism exerts a strong appeal for the anxious voter. Populist messages rely on simplistic answers to complex problems and promote an us-versus-them warfare mentality. Like Mr. Trump, populists engaged in battle have traditionally ridiculed their opposition; but in the narcissistic endeavor to prove himself a winner at the expense of all those "losers," Trump relies on righteous indignation, blame, and contempt as weapons of war. Many disaffected voters are drawn to him precisely because of those traits and not in spite of them.

HUCKSTERISM

The term *huckster* refers to a person who peddles something purely for self-interest or financial gain, using showy tactics and hyperbolic language to perpetrate his scheme on others. The huckster speaks the language of truthful hyperbole, knowing full well that he can manipulate people's minds with verbal bluster and braggadocio. Historically, the term *huckster* was applied to any type of vendor or seller, but over time it has come to assume distinctive negative connotations referring to con artists, hustlers, and swindlers. History and literature are replete with such personages. Shakespeare's plays, for instance, are full of shady characters who make do by perpetrating half-truths, making false promises, and duping others to follow them or to give them what they want. One of the most famous of all huckster archetypes in literature and the arts is the *Dottore* ("Doctor"), a stock character of the *Commedia dell'arte*, who was hyperbolically loquacious, hiding his pedantry cleverly with untruths and alternative facts. He always had a medicine or potion at hand that he could sell to cure any condition possible, from itchiness to marital infidelity. Significantly, the Dottore is a pompous Latin-spouting pseudo-scholar, who actually uses malapropisms and gibberish. He cannot open his mouth without spitting out bluster. He is a huckster who is a master at truthful hyperbole.

In America, the figure of the huckster is akin to the figure of the wheeler-dealer con artist. One of the first appearances of the American con man is in Herman Melville's 1857 novel *The Confidence-Man: His Masquerade*,[29] imprinting it into the American imagination permanently. Outraged at the plethora of con men in America, but especially in business and politics, Melville wrote his novel as a cautionary tale about the social destruction that ensues when truthful hyperbole becomes the basis of dealings among people. It recounts what happens when the devil boards a riverboat traveling down the Mississippi River. He goes unrecognized as the "evil one" because he is dressed in disguise, boarding the vessel to conduct shady business deals as strategies for perpetrating evil. In a culture of greed and rampant materialism, the con man is never recognized for who he really is, Melville suggests. His false claims are readily believed because of the false monetary promises that he makes. The huckster is, as Melville certainly understood, a master of the Art of the Lie, using guile, cunning, deviousness, and slyness to dupe ingenuous and naïve people to do his bidding, because they too are greedy and anxious to gain success at any cost. The huckster poses a real danger to the American Dream, which is built on honesty, altruism, and truth. The moral of Melville's story is that the con man will eventually destroy America, since he has the keen ability to manipulate victims by gaining their confidence through hyperbolic aplomb and the skillful telling of lies. Once entrapped by the huckster's web of lies and falsehoods, we are all inclined to give him our trust. Mark Twain was also keenly aware of the dangers of con men and huckster scoundrels to American society, interspersing such characters in his novels. Similar figures are found in George Ade's 1896 novel, *Artie*, whom he describes as having the ability to deceive others with their "large, juicy con talk."[30]

Trump is the quintessential American huckster, eerily resembling the con men in Melville's, Twain's, and Ade's novels in the way that he presents himself and through his deceptive schemes, which he presents with his "large, juicy talk." In a truly perceptive article, Max Boot describes Trump in this way, showing how he has literally pulled the wool over people's eyes, like a true huckster of American folklore.[31] Like Burgo above, Boot asks, however, if this truly matters to his fans, who have accepted him as their deal-making hero: "Voters knew what sort of huckster Trump was when they elected him. But it should give us pause to consider what it says about America, circa 2018, that so many of us are so ready to accept . . . a con man . . . as our leader."[32] Hannah Arendt, who escaped Nazi Germany, would often point out how dangerous this kind of person and his bombastic speech are, because he knows how to avoid criticism and because he never tells the truth, using the big-lie technique, and thus coming across as believable.[33] As American journalist and essayist Walter Lippmann argued in his 1922 book, *Public Opinion*, the language of the marketplace huckster gains people's trust because it produces "pictures in our heads" of promises that cannot be attained easily in reality.[34]

Trump can be described as a contemporary Jack Dawkins—the "Artful Dodger" in Charles Dickens's 1837 novel *Oliver Twist*.[35] Dawkins is portrayed as a skilled pickpocket who uses cunning and artful mendacity to ply his criminal trade. Like Trump he is able to finagle his way out of dangerous predicaments by using his gift of the gab and his keen sense of how to bullshit (an expletive that will be discussed subsequently). In Dickens's novel, Dawkins portrays himself as a "victim of society," but he ends up in jail nonetheless. Trump too portrays himself as a victim of the deep state. And like Dawkins, he too comes across as a loveable scoundrel to his followers—but a scoundrel nonetheless.

There have been many Jack Dawkinses and Donald Trumps in America's history, having become legends of sorts. They were masters of disguise and shiftiness, able to get away with their lies, deceit, and tricks via their oratorical wizardry, an amalgam of big talk and big lies. Perhaps the most important lesson to be learned from the Trump presidency is that it exposes how we are all susceptible to con games, as Melville so obviously understood. Machiavelli was similarly keenly aware that hucksterism works, because we are all gullible to deceit, using the example of Alexander the Sixth, one of the most controversial of all popes, who held on to power through deceit and a Renaissance version of hucksterism:[36]

> Men are so simple, and so subject to present necessities, that he who seeks to deceive will always find someone who will allow himself to be deceived. One recent example I cannot pass over in silence. Alexander the Sixth did nothing else but deceive men, nor ever thought of doing otherwise, and he always found victims; for there never was a man who had greater power in asserting, or who with greater oaths would affirm a thing, yet would observe it less; nevertheless his deceits always suc-

ceeded according to his wishes, because he well understood this side of mankind.

Trump's con game is so ingenious that he has convinced many to read him, literally, as an atypical businessman-politician-preacher who runs the country like a corporation. The Art of the Lie in a huckster's hand can be easily transformed into the Art of the Con. And like a clever huckster, Trump himself accuses others of being cons, over and over. This is the primary skill of hucksterism, deflecting attention away from oneself, as Ben Zimmer has perceptively pointed out in an article for *The Atlantic*; it is a textbook case of projection:[37]

> But perhaps the most significant word of all for Trump was a three-letter one: *con*. More than a dozen times he used it in the phrases *con artist, con job*, or *con game*. First, he called out Democrats as "con artists" for destroying the reputation of his Supreme Court pick, Brett Kavanaugh, characterizing the mounting accusations of sexual misconduct as "a big fat con job." The lawyer Michael Avenatti, representing one of the accusers, got singled out as a "con artist," and he insisted that even "George Washington would be voted against 100 percent by [Senator Chuck] Schumer and the con artists."

Linguist David Maurer wrote a truly perceptive book back in 1940, titled appropriately *The Big Con*, in which he gave the first full description of the features and effects of the big talk of hucksters and how it renders us credulous despite evidence that we are being conned.[38] Maurer's book inspired the 1973 movie *The Sting*,[39] which is a portrait of American hucksterism and how it has become an intrinsic part of American culture. Actually, even before *The Sting*, Hollywood movies in the 1940s provide implicit psychological portraits of hucksters and their negative effects on society. For example, the 1947 movie *The Hucksters*[40] is about a New York adman who has the gift of the gab and an uncanny ability to come up with the right slogan for a product. His duties take him to Hollywood, where he creates a successful radio commercial for "Beautee Soap," which mimics the jingle style of the era. The movie emphasizes how ad agencies control what people see, shaping social values more than writers and artists. In *A Letter to Three Wives* (1949),[41] Ann Sothern is a writer of radio soap operas. Her husband is a critic of the media and advertising worlds, which he denounces as vulgarizing American culture. In one scene, the husband states: "The purpose of radio writing, as far as I can see, is to prove to the masses that a deodorant can bring happiness, a mouthwash guarantee success and a laxative attract romance." In the 1957 film *A Face in the Crowd*,[42] Andy Griffith plays a homeless person who is hired to act in commercials because of his ability to charm consumers by poking fun at the sponsors of programs. The movie constitutes a black parody of advertising culture and its diffusion into politics and society.

The subtext in these movies is that the huckster may appear to be a charmer on the surface, but he surreptitiously destroys the moral fiber of a society. Philosopher Max Black wrote perceptively in 1982 about hucksterism and lying as follows:[43]

> [A] familiar observation [is] that the liar is parasitic on general, though not universal, veracity: lying, as a species of deceit, would be futile in the absence of general efforts to be truthful. It seems reasonable to conclude that a liar is, in a radical way, sapping the foundations of social institutions, all of which depend on the general effectiveness of speech. The liar is indeed an enemy of society, who tends to undermine all possibility of civilized intercourse. Universal lying would destroy intelligible speech.

BULLSHITTING

The huckster knows not only how to "talk big" about anything to anyone, but also how to talk his way out of difficulties and challenges. Colloquially, this is called "bullshitting," an expletive that I cannot avoid since it describes colorfully a type of strategy that is essential to the enactment of the Art of the Lie. It can be defined as skill at making up things on the spot in order to either extricate oneself from some difficult predicament or else to present information to support something fallacious through shifty, evasive language. One possible origin for the term is, according to literary scholars, T. S. Eliot's unpublished poem "The Triumph of Bullshit,"[44] in which the word appears only in the title. Eliot's coinage of the term was probably intended as an indirect attack on some of his critics, using an excremental image to do so. But whatever the origin of the word, it has now become a common one for characterizing the kinds of evasive mannerisms of speech that hucksters, con artists, and other shady people utilize to bamboozle their victims and shield themselves from being exposed as fraudulent.

There are so many examples of bullshitting perpetrated by Trump that it would take a huge tome simply to list them. He has even been nicknamed the "Bullshitter-in-Chief." Bullshitting is effective because, on the spot, people do not generally challenge the bullshitter, or when they do, he can simply postpone the challenge with "We'll see," in the knowledge that the confrontation will likely dissipate over time. Trump's claim that, contrary to what his intelligence agencies have maintained, Russia never influenced the election has been shown to be knowingly deceptive. So, when challenged, he typically points out that the evidence to support his claim "is coming," relying on his acolytes to bring forth (false) evidence to support him. Over time, the purported evidence predictably evaporates into the political ether. Trump claimed that he won more electoral college support than anyone else in the past. When a reporter called Trump out for spreading this easily verified false information, he simply said, "Well, I was just given that information. I don't know. I was just given it. We had a very big margin."[45] Similarly, his ludicrous claim that "thousands" of illegal voters were bussed in from Massachusetts to vote in New Hampshire was pure bullshit. No such evidence has ever surfaced to support this bogus claim.

For bullshitting to be effective, it must appear to be truthful, and this involves skillful deceptive performances of the falsehoods. That is to say, the actual delivery of the falsehood must be convincing and decisive, precluding any counterattack or rebuke. It is the same tactic used by schoolyard bullies, especially when they have cohorts and allies willing to back them up. The delivery is a bald-faced fake performance, usually strengthened by claims of expertise, with no requirement to validate them ("I know best"). The objective is, of course, to postpone the day of reckoning, allowing it to evanesce into the fog of ever-changing news cycles—a saliently obvious ploy that can be seen in Trump's constant use of "we'll see," "people tell me," "you know that too," or "you understand that." In a speech Trump gave at MacDill Air Force Base in Tampa, Florida, in February 2017, he claimed, with no proof whatsoever, that:

> All across Europe you've seen what happened in Paris and Nice. All over Europe it's happening. It's gotten to a point where it's not even being reported. And in many cases the very, very dishonest press doesn't want to report it. They have their reasons, and you understand that.[46]

In a cogently argued book, *Post-Truth: How Bullshit Conquered the World*, journalist James Ball has documented the deleterious effects of bullshitting on society, since, as he suggests, it is upsetting the emotional balance needed for a society to thrive.[47] While we know that a statement such as the one above is patently incorrect, it still lingers in our minds. Bullshitting is thus not only an effective deflective strategy that obviates admitting the truth when such admittance would jeopardize the reputation of the politician, but also a mental health issue (as discussed).

One of the most insightful and comprehensive treatments of bullshitting is the one by Harry G. Frankfurt, *On Bullshit*, in which he distinguishes between an outright lie and bullshit as follows:[48]

> Someone who lies and someone who tells the truth are playing on opposite sides, so to speak, in the same game. Each responds to the

> facts as he understands them, although the response of one is guided by the authority of the truth, while the response of the other defies that authority, and refuses to meet its demands. The bullshitter ignores these demands altogether. He does not reject the authority of the truth, as the liar does, and oppose himself to it. He pays no attention to it at all. By virtue of this, bullshit is a greater enemy of the truth than lies are.

Frankfurt references a key 1982 paper by philosopher Max Black (mentioned above), who uses the term *humbug* in place of bullshit, characterizing it as follows:[49]

> Humbug has the peculiar property of being always committed by others, never by oneself. This is one reason why it is universally condemned. No doubt we can agree that humbug is a Bad Thing; but what are we agreeing about? It proves astonishingly hard to say. . . . We have already seen that violations of the communicative framework need not consist in the utterance of falsehoods. If I reply on the telephone to the question "Have you got any sausages today?" by saying, "No," and continue in the same vein, saying that I won't have any in the foreseeable future, and the like, everything I say may be literally true, but I shall deceive the other as if I were deliberately lying.

Trump has all the skills of the master bullshitter—skills described insightfully by Stanton Peele and worth paraphrasing here:[50]

> The bullshitter knows that people are normally afraid to challenge him, because the confrontation would violate rules of social politeness, and because they might fear that their own illicit or fraudulent actions will be found out.
>
> The bullshitter emphasizes his legitimate successes, laying the psychological groundwork for gaining credibility. By putting his name on his buildings, Trump is able to emphasize his past successes, allowing him

> to promote any bullshit that he sees as advantageous when discussing his redesign of the polity.
>
> The bullshitter is a Machiavellian lion (hiding the fox within), constantly coming across as forceful, which is an intimidation technique. Acting and speaking modestly is a sign of weakness. On the other hand, arrogantly attacking one's accusers is a sign of strength.
>
> Claiming esoteric knowledge is another key ploy in the bullshitter's arsenal of tricks. Trump constantly claims that he "knows more" than anyone else in any field, from the military to climate science. It is bullshit, but it works because he presents it with a don't-question-me tone.

The foregoing discussion brings us back to the notion that we are living in a "post-truth" era (discussed previously). The assumption is that there was an ideal era when truth reigned in all official matters, or at least was enforced through legislation and social censure. This is the legacy of the Enlightenment and its roots in Socratic classicism. In Machiavelli's Renaissance, what mattered was the "appearance of truth," not actual truth, in the conduct of political and official matters. Life went on just the same. It was common knowledge that rulers, clerics, and the authorities lied, even more effectively than common folk. Bullshitting is actually part of an "art of making do," as Joseph Pine has so cogently argued.[51] This inheres in a social *ethos* that accepts bullshitting as an unconscious principle of social interaction, making it relatively harmless if this is known by people. The art of making do is, in effect, a residue of the ancient Sophist philosophy of the world, where cleverness and fallaciousness are perceived to be part of normal dialogue and, indirectly, a means of fleshing out the truth through counterargumentation. The Enlightenment aimed to eliminate such strategies and restore Socratic truth as the only kind of speech. So it is perhaps more correct to say that we live in a post-Enlightenment era, rather than post-truth one, defined by hucksterism and bullshitting. We live in an age where the bullshitter is given as much social space as the truth teller, and indeed the latter may even come across as somewhat foolish. A brief digressive foray into the Enlightenment mindset might be useful at this point.

After the voyages to the Americas in the late fifteenth century, there arose a heated debate about ethnicity throughout European academies and societies broadly. Late in the sixteenth century, the French essayist Michel de Montaigne tried to dispel the derogatory popular view that had arisen in Europe with respect to the indigenous peoples of the Americas, arguing that their cultures were as valid as European ones—just different adaptations to the world.[52] It was important to understand these cultures in human terms, not in terms of European worldviews. But Montaigne's reasonable viewpoint had to await the eighteenth century to gain acceptance and currency. In that century, the Age of Enlightenment proclaimed the view that all cultures should be studied rationally in the same way that one studied physical nature, and that no one culture was superior to any other. Science and logic would dispel all bias and bigotry, and would answer the broader human philosophical questions, without resource to superstition and mythology. Only in this way was truth achievable, detached from belief and dogma. Enlightenment intellectuals reexamined and questioned all received ideas and values, exploring new ways of thinking rationally. The Enlightenment marked a pivotal stage in the growth of modern secularism and the objective study of cultures and races. It led to the foundation of humanistic sciences such as anthropology and historiography. In the type of society envisioned by the Enlightenment, lies would be rejected outright as anomalies of human psychology.

Belief systems were seen as important for social cohesion, but subsidiary to the ideals of science and its basis in logical reasoning. Coupled with the advent of Darwinian evolutionary theory, the stage was set to eliminate superstitions and the many lies that these produced. The Enlightenment stressed the need for individual rights and, more importantly, individual critical thinking, free from the yoke of credulous thinking. But the movement forgot, or ignored, that humans lie for a host of reasons and have always done so, as Dallas Denery has so comprehensively documented.[53] In his *Discourse on the Origins of Inequality*,[54] Jean-Jacques Rousseau—an Enlightenment philosopher—claimed that the origin of lying and deception can actually be found in human evolution, suggesting that humans, who were at one time solitary wanderers, came together to form the first families and societies. As they developed agriculture, the need to divide land emerged, and from this they started to use language deceptively to gain advantage, for reasons of territoriality. So, lying in this framework is a survival mechanism. Now, Rousseau's theory might be seen as somewhat fanciful, but there is a grain of truth in it. It might well be that we developed artful strategies to outwit our opponents for reasons of survival and then these became imprinted in the collective unconscious as general strategies. On the other hand, lying may be the product of human ingenuity, which can be used for bad and good. Whatever the case, our so-called post-truth era is not unique; the quest for truth has always been impeded by mendacity and falsehoods.

EPILOGUE

Bombastic, hyberbolic language assails people like a runaway train coming directly at them. There is little that can be done to stop the train. It plays on the belief that anyone can become successful no matter what humble background one may come from. The language of truthful hyperbole can be compared to a vitality tonic for the mind, imbuing it with a feeling of vigor and strength. It is an effective discourse strategy for Machiavellian liars, since it gives them the power to penetrate the mind and feed on it lethally. As Robert Louis Stevenson so aptly put it, "The human being is a creature who lives not upon bread alone, but principally by catchwords."[55]

The language of truthful hyperbole is a direct descendent of P. T. Barnum's businessman-preacher oratorical style. It reduces discourse to formulas, stock phrases, jingles, and slogans. It is useful to recall here Vance Packard's 1957 indictment of this kind of language as a surreptitious form of persuasion, in his widely read work *The Hidden Persuaders*.[56] Sadly, the language of truthful hyperbole is fast becoming the koiné of cyberspace. As James B. Twitchell aptly put it a while back, "Language about products and services has pretty much replaced language about all other subjects."[57] Many people react to the language of truthful hyperbole in ways that parallel how individuals and groups have responded in the past to religious oratory. It has become a ubiquitous, all-encompassing form of discourse. Since the end of the nineteenth century, this kind of verbal wizardry has succeeded, more so than any economic process or socio-political movement, in promoting and ensconcing consumerist lifestyles as the only important ones. Stuart Ewen puts it eloquently in the following manner:[58]

> If the "life-style" of style is not realizable in life, it is nevertheless the most constantly available lexicon from which many of us draw the visual grammar of our lives. It is a behavioral model that is closely interwoven with modern patterns of survival and desire. It is a hard to define but easy to recognize element in our current history.

The post-truth era is an oppressive one, shaped by a morbid dependence on technology, as Bob Stein has discerningly remarked:[59]

> When 1984 came and went, Americans congratulated themselves on the fact that Orwell's Big Brother had not materialized in the West. But what people missed, of course, was that Huxley's infinitely darker vision had come true . . . In *Brave New World*, Huxley saw a time coming when "people will come to love their oppression, to adore technologies that undo their capacities to think."

Sardonically, Trump proclaimed that the "American Dream Is Dead"[60] in his announcement that he was running for the presidency in 2015, indicating that he would make America great by putting "America First"—a hyperbolic statement that became a mantra for his election. In a thorough assessment of his use of this hyperbole, literature professor Sarah Churchill argues that Trump's slogan, unlike the American Dream's promise of equality for all, was an early slogan of the Ku Klux Klan.[61] Significantly, she observes that a Klan riot in 1927, which led to the arrest of seven men, included Donald Trump's father—Fred C. Trump. The America First slogan, she suggests, is not a true patriotic one; it is white supremacist wordplay.

As the nineteenth-century writers certainly knew, modern America was, sadly, shaped by hucksters like P. T. Barnum as much as it was by dreamers. In Timothy O'Brien's biography of Trump, he cites his sister as saying that her brother "is P. T. Barnum," a big talker with no real substance to his talk other than self-promotion.[62] Trump became a television star specifically to promote himself as an impresario in true American huckster fashion, providing illusory promises to anyone. People visiting Barnum's exhibitions did not care that what they saw there was make-believe rather than archeological fact. The desire to immerse themselves in a world of fantasy filtered their sense of reality and enhanced their believability. All one has to do, as Barnum knew, is to promise dreams (especially unattainable ones) and people will flock to your circus.

8

A MACHIAVELLIAN ART

The only thing worse than a liar is a liar that's also a hypocrite!

—Tennessee Williams (1911–1983)

PROLOGUE

Since antiquity we have been fascinated by stories of deceit, betrayal, and cunning. We sense that mendacity is a trait of the human brain. Psychologists have given this trait, as discussed previously, a name—*Machiavellian Intelligence*, a term introduced by Frans de Waal's widely cited 1982 book, *Chimpanzee Politics*, in which he refers to Machiavelli to support his view that lying emerged as a social characteristic.[1] De Waal's main claim is that lying enhances the ability of the individual to control and manipulate social interaction advantageously. Therefore, to interpret this claim in terms of the present discussion, it would seem that those who were better at lying would be more successful in social competition. This might explain why Machiavellian liars have the ability to manipulate others so successfully—we may be predisposed by our evolution to see them as having leadership qualities.

Although it has been critiqued on many fronts, de Waal's derivative idea that political success is dependent on the skillful use of mendacity is a plausible one, as has been discussed throughout this book. This implies that some are more "talented" than others at using lies to promote themselves by gaining trust through their ability to influence people with words.

While we all lie from time to time in order to gain some advantage or to avoid unwanted consequences, few of us are masters of the lie. The latter are astute "readers" of others, understanding how to get into their minds and manipulate them for personal gain. They are "artists" of manipulation, keenly aware of what words can do to influence thoughts. The goal of this final chapter is to argue that mendacity may, or may not, be an evolutionary trait, but in its extreme forms it has hardly helped humanity. Machiavellian liars are narcissists who do not contribute to the progress of humanity, but instead put obstacles in its thrust forward, as the history of dictatorships has shown. The master liar has the ability to mesmerize people, prodding them to do what he wants them to do. He exerts a charismatic magnetism, like a cult leader, inducing a kind of hypnotic trance in people that is hard for them to shake off. It was sociologist Max Weber who introduced the idea that cults were based on the psychology of charisma:[2]

> Charisma is a certain quality of an individual personality by virtue of which he is set apart from ordinary men and treated as endowed with supernatural, superhuman, or at least specifically exceptional powers or qualities. These are such as are not accessible to the ordinary person, but are regarded as of divine origin or as exemplary, and on the basis of them the individual concerned is treated as a leader.

Like other Machiavellian liars before him, Trump mesmerizes his followers with his particular brand of Barnum-type charisma, making them believe that he is a victim of their own purported enemies, who have perpetrated atrocities on them. As Orwell so aptly put it: "Everyone believes in the atrocities of the enemy and disbelieves in those of his own side."[3]

The Art of the Lie is a Machiavellian art. It is the art of hucksters and con artists, as Herman Melville and Mark Twain warned. Perhaps at no other time in human history as the present one has it become so destructive. In a world of algorithms and memes, the difference between truth and lies is blurry. Fortunately, the lies and actions of a Donald Trump, among many other opportunistic liars, have mobilized many to bring about change. The battle for truth is an ongoing one, as it has always been throughout our history. Lying may be in our genes, as de Waal's work suggests, but the human imagination can overcome its deleterious effects with will power—that has been one of the most important of all lessons to be learned from human history.

MACHIAVELLIANISM

The term *Machiavellianism* refers to the type of person who espouses a deceitful or duplicitous style of speech, possessing a cynical disregard for others. Those who use such language advantageously have been termed master liar-princes in this book. It was Machiavelli who provided the first handbook for liar-princes, in which he put forth strategies on how to lie, deceive, and confabulate effectively. In fairness, Machiavelli preferred "free republics" to principalities and, thus, to government determined by the citizenry instead of government by a single ruler. But he also understood that power is rarely gained by honesty; cleverness and cunning are better strategies.

As discussed throughout this book, Machiavelli described the master liar as both a fox and a lion—a cunning manipulator of words who must always appear to be fearsome and powerful in order to defeat the "wolves" (his opponents). The liar-prince must thus "be the fox to avoid the snares, and a lion to overwhelm the wolves."[4] Significantly, the Catholic Church banned *The Prince* in 1559, seeing it as an immoral treatise. But it resurfaced thereafter as a widely read book, remaining a controversial work to this day.

The influence of Machiavelli's *The Prince* on history cannot be underestimated. The St. Bartholomew's Day Massacre of 1572 in Paris—a series of assassinations against the Huguenots (Calvinist Protestants), believed to have been prompted by Queen Catherine de' Medici, the mother of King Charles IX—was attributed by Innocent Gentillet (a Huguenot) to Machiavellianism in his 1576 book, *Discours contre Machievel*.[5] In *Henry VI, Part III*, Shakespeare describes the Machiavellian liar as a "chameleon" who changes "shapes with Proteus for advantages."[6] French philosopher Denis Diderot described Machiavellianism as the "art of tyranny,"[7] whose sole purpose was to deceive and manipulate others, going against every and all conventional moral codes of humanity.

In a fascinating 1970 book, Richard Christie and Florence L. Geis developed a "Machiavelli Test" for measuring the level of Machiavellianism in people.[8] Those who scored high on the test tended to endorse a statement such as the following one: "Never tell anyone the real reason you did something unless it is useful to do so." The results of the study suggest that Machiavellianism might be present to varying degrees in all of us, as de Waal suspected, varying in its manifestations according to subject and context. The main features of Machiavellianism have, since this key work, been studied in some detail.[9] These include the following:

- focusing on ambitions and self-interests;
- seeing money and power as more important than relationships;
- knowing how to effectively exploit and manipulate others to get ahead;
- using lies and deception whenever required;
- causing others harm to achieve one's own ends;
- possessing very low levels of empathy;
- never revealing one's true intentions;
- reading social situations and others perspicaciously.

In some psychological circles, Machiavellianism is considered to be part of a so-called Dark Triad, with the other two being narcissism and psychopathy.[10] As psychologist Glenn Geher has written, Trump rates high on the Dark Triad scale:[11]

Many reach the top by being conspicuously caring—demonstrating a lifelong dedication to their broader communities and to helping others in their social worlds. Think Mother Theresa. On the other hand, there are relatively dark ways to reach the top in nearly all human social contexts. Displaying characteristics of the Dark Triad—being uncaring about others, self-absorbed, and manipulative—for better or worse, seems to also be an effective route to the top. It may not be a nice approach to social life, but it can be a successful one—particularly if others in the community allow this kind of strategy to succeed. Does Donald Trump demonstrate the features of the Dark Triad? Based on my expert opinion having published extensively in this area of psychology, I think the answer is this: Absolutely and unequivocally.

LIES AND MENTAL HEALTH

In the previous chapter, the connection between lying and mental health was discussed briefly under the rubric of "words matter." It is worthwhile to revisit this topic here since a main claim of this book is that lying might be destructive of mental health.

We hardly realize how much language and culture affect disease and our medical approaches to it. In Western languages, metaphors of pain and disease reveal that we perceive the body as a machine ("My body is not working today"; "It is shutting down"; etc.). These verbal formulas predispose us to conceptualize pain as a malfunction in the machine (the body). This linguistic-cultural model has guided the view that pain is something that can be detected and eliminated, by correcting defects in the machine, so to speak. This model goes back to Jacques de la Mettrie's 1747 book, *L'homme machine*, which characterizes medical techniques as "repairs" to the biological machinery.[12] As a model, it has produced many positive results (needless to say). But the same view cannot be assumed cross-culturally. Speakers of Tagalog, for example, have no equivalents of the previous expressions. Their words reveal instead that body health is influenced by spiritual and natural forces. These two different patterns of groupthink produce different responses to pain and disease. People reared in English-speaking cultures are inclined to experience pain as a localized phenomenon; that is, as a malfunction that can be corrected apart from the overall state of well-being of the individual. Tagalog speakers, on the other hand, are inclined to experience it as intertwined holistically with mental states and contextual factors and, therefore, as connected to the overall state of well-being of the person.

Jacalyn Duffin has argued that throughout history illness is often what we define it to be.[13] She points out that "lovesickness" was once considered a true disease, even though it originated in the poetry of antiquity and was reinforced in the poetry of the medieval period. Its elimination as a disease from medical practice is due to twentieth-century skepticism and scientific research, which finally exposed it as a cultural construct. Her point is that, at any given point in time, concepts of what constitutes a disease might crystallize from beliefs, not science. These then affect how we experience the disease. The implication is that our constructs might influence our health adversely, and that these are encoded primarily in language.

It is no stretch to say, therefore, that lies might affect mental health. A relevant study by Anita Kelly and Lijuan Wang, found that Americans average around eleven lies per week. On the basis of 110 subjects over ten weeks, they also found that half of the participants who were instructed to stop telling lies showed significant mental health improvements.[14] Without going into details here, studies such as this one have started to show that the parts of the brain linked to the emotions are affected detrimentally by lies.

One of the first anthropologists to be aware of the deleterious effect that lying may have on mental health was Gregory Bateson, who put forth the hypothesis that schizophrenia has to do, in part, with a difficulty in distinguishing aspects of meaning.[15] This is the basis of Bateson's theory of the double bind, which refers to a communicative situation characterized by recurring paradoxical messages: a primary injunction ("Do not do this") and a secondary one ("Do not see this as a punishment"). The schizophrenic has difficulty discriminating the meanings of such messages. This has therapeutic implications according to Bateson: the family of a schizophrenic patient must avoid double binds, which produce negative thoughts. If we change Bateson's term *double bind* to *doublespeak*, the more general implications for mental health that his theory might imply come into focus.

MANIPULATIVE LANGUAGE

Manipulative language is harmful to mental health because it stokes feelings and beliefs that affect emotional well-being. We are susceptible to it. As mentioned in the previous chapter, Machiavelli gave the example of Pope Alexander VI as the master manipulator, because he used lies and deception throughout his papacy, showing that people are easily manipulated by words:[16]

> Alexander the Sixth did nothing else but deceive men, nor ever thought of doing otherwise, and he always found victims; for there never was a man who had greater power in asserting, or who with greater oaths would affirm a thing, yet would observe it less; nevertheless his deceits always succeeded according to his wishes, because he well understood this side of mankind.

Each time we hear or read words, images crop up in the mind. By simply uttering the word *cat*, people understand what is being singled out in reality, even though an actual "cat" may not be present to observe with our sensory system. Similarly, by simply saying the word *minotaur*, we will understand what is being implied, even though no such thing is available for the senses to detect. It is not real in any concrete sense; it is a construct that comes from mythic stories. The remarkable feature of language is its ability to conjure up anything at will, even if it is not real but imaginary. When an image is produced by a word, there is virtually no way to eradicate it from the mind. Manipulating images is the essence of the Art of the Lie. When Trump iterated "Crooked Hillary" over and over, those who were disgruntled at the Obama administration gladly accepted this as a truth, even if it was not justifiable. It stoked an image which, like the *minotaur* one, exists in the imagination.

The idea that language shapes people's perception of reality and that lying can alter that perception caught the attention of the Gestalt psychologists in the 1920s and 1930s. For example, Carmichael, Hogan, and Walter conducted a truly remarkable experiment in 1932.[17] They found that when they showed subjects a picture and then asked them later to reproduce it, the drawings were influenced by the verbal label assigned to the picture. The picture of two circles joined by a straight line, for instance, was generally reproduced as something resembling "eyeglasses" by those subjects who were shown the *eyeglasses* label. On the other hand, those who were shown the *dumbbells* label tended to reproduce it as something resembling "dumbbells." There is really no other way to explain these results, other than by the fact that language conditions the way we see things in our minds.

Ann Gill eloquently describes the link between words, thought, and actions as follows:[18]

> By portraying experience in a particular way, words work their unconscious magic on humans, making them see, for example, products as necessary for success or creating distinctions between better or worse—be it body shape, hair style, or brand of blue jeans. Words create belief in religions, governments, and art forms; they create allegiances to football teams, politicians, movie stars, and certain brands of beer. Words are the windows of our own souls and to the world beyond our fingertips. Their essential persuasive efficacy works its magic on every person in every society.

The Machiavellian liar understands that the brain is a malleable organ that can be altered by mendacity; that is, by manipulating the link between words and thought. The triumph of a Mussolini, a Hitler, or a Trump rests on a masterful manipulation of language, more than it does on anything else.

The ancient orators were well aware of the power of language to control minds, establishing the art of oratory to study how this power could be harnessed. They described effective oratory in terms of five strategies. The first one was *inventio* (invention), or the search for an argument that will gain attention. The birther story is a classic example of *inventio* for nefarious reasons. With it, Trump was able to stoke resentment in those who saw the Obama years as "un-American." The made-up story of being born outside of America and being of Muslim descent also fit in cunningly with the false narrative that Muslims were dangerous. The Machiavellian liar clearly knows how to tap into beliefs, bringing them out in the open through *inventio*.

The second strategy is *dispositio* (arrangement), which is the organization of the speech act to make it convincing. Words gain resonance when they are put together into utterances that resonate with interlocutors. Trump's use of truthful hyperbole is a *dispositio* tactic, since it allows him to put forth promises without providing any proof that he can realize them. The third device is *elocutio* (style), which relates to how the delivery of speech should unfold with maximum efficacy. This has always been a major tactic in how Trump has fashioned his persona. At his rallies, he uses derision and ironic attacks on opponents, calling them names or slurring their reputation. He is a master at *elocutio*.

The fourth tactic is *memoria* (memory), which implies that the orator should talk about things that evoke specific memories, thus getting his interlocutors to believe that he knows them personally, or that he is "one of them" who "understands" their plight. When Trump promised to make everyone wealthy during the campaign, claiming he "understood" the plight of factory workers, coal miners, and other blue-collar workers, even though he never worked in any factory or coal mine, he was playing on *memoria*, and the sense of historical injustices that his audience members felt previous liberal politics had foisted on them. He also played on the fears in his audience members of losing their jobs to outsiders, claiming to "bring back the jobs" to America, evoking images of an "invasion" of foreigners to America stealing jobs away from the denizens of the nation.

Finally, *actio* (delivery) is the tactic of making the delivery of speech effective by knowing one's audience and adapting to it. This aspect of Trump's speeches and overall language has been discussed throughout this book and needs no further elaboration here.

The ancients also understood that the antidote to manipulative oratory is oratory that promotes truth. This is an effective strategy, as attested by the great speechmakers of history, from Cicero to Martin Luther King Jr. As the philosopher Parmenides (c. 450 BCE) implored his contemporaries to do in his poem, "The Way of Truth,"[19] the quest for truth leads us to those things that are unchangeable over time.

COGNITIVE DISSONANCE

When believers of conspiracy theories, confabulations, and other forms of doublespeak are faced with evidence that these are baseless, they tend to experience what psychologists call "cognitive dissonance"—a term introduced initially by Leon Festinger in 1957.[20] Cognitive dissonance results from a discord between one's beliefs and facts. To resolve the dissonance, people will seek out information that confirms their false beliefs, rather than reject them, avoiding information that is likely to be in conflict with them. Festinger found that people develop strategies that are designed to attenuate the dissonance and even turn the contrasting information on its head to make it fit their beliefs. For this reason, it is unlikely that people who have been manipulated by a master liar will ever change their minds about anything he says. In his book *When Prophecy Fails* (written with Henry W. Riecken and Stanley Schachter), Festinger puts it as follows:[21]

> A man with a conviction is a hard man to change. Tell him you disagree and he turns away. Show him the facts or figures and he questions your sources. Appeal to logic and he fails to see your point.

The need to resolve cognitive dissonance may be a primary reason why followers of a despot tend to stay with him virtually to the end. People who are duped by con artists, hucksters, or shysters will often remain tacit or incapable of taking action against them after discovering that they were conned. Once a consummate liar is believed, it is almost impossible to see through his lies and to accept the truth. The feeling of being duped is so destructive emotionally that it is easier to deny the facts or else explain them away in some self-illusory way than to accept them. In the case of someone who has become invested emotionally in the liar because of deeply held beliefs that he has stoked, any proof provided against him will actually impel the deceived person to dig in even more, as Festinger, Riecken, and Schachter found:[22]

> Suppose an individual believes something with his whole heart; suppose further that he has a commitment to this belief and he has taken irrevocable actions because of it; finally, suppose that he is presented with evidence, unequivocal and undeniable evidence, that his belief is wrong: what will happen? The individual will frequently emerge, not only unshaken, but even more convinced of the truth of his beliefs than ever before. Indeed, he may even show a new fervor for convincing and converting other people to his view.

The theory of cognitive dissonance might explain why the great despots of history have been able to get so many to submit to their will, even at the cost of their own personal liberty. As Mussolini so aptly put it, the underlying goal of any effective leader is to eliminate "the putrid corpse of liberty."[23]

Cyberspace has become, increasingly, a place that produces cognitive dissonance on a daily basis—a space where the distinction between truth and mendacity is a tenuous one. It has produced a state of mind that is constantly involved in resolving the cognitive dissonance that this lack of distinction entails. In the global village, we are caught in the present, making us feel part of the "speed of light" in which we are constantly immersed, as Marshall McLuhan observed: "At the speed of light, there is no sequence; everything happens in the same instant."[24]

The world in which we live is, in a phrase, a cognitively dissonant one. This is likely to be a major consequence of living in a *technopoly*, a term coined by Neil Postman in 1992, in his book *Technopoly: The Surrender of Culture to Technology*.[25] Postman defines a technopoly as a society that has become pathologically reliant on technology, seeking authorization in it, deriving recreation from it, and even taking its orders from it.

In a technopoly, what counts is information itself, not if it is true or false. This is a coping strategy. Postman posits three historical phases whereby technology and human evolution can be seen to dovetail. The earliest phase, which he designates as *tool-using*, is an era in which tools are invented and used to solve physical problems of survival and to serve ritual symbolism and art. These cultures are theocratic and unified by a metaphysical view of the world. He calls the second phase *technocratic*, an era in which tools are connected to a particular worldview or "thought-world," as he terms it. This era overthrows the previous metaphysical thought-world. Technocratic cultures impel people to invent, hence the rise of science and literacy. Finally, a *technopoly* is a "totalitarian technocracy," evolving on its own. It reduces humans to seeking meaning in machines. In this environment, the search for objective truth is no longer felt to be a value; only information itself has value. In this state of mind, cognitive dissonance is constantly at work, inducing people to resolve conflicts between fact and fiction, truth and lies, simulation and reality. As engineer Jaron Lanier warns:[26]

> We [engineers] make up extensions to your being, like remote eyes and ears (webcams and mobile phones) and expanded memory (the world of details you can search for online). These become the structures by which you connect to the world and other people. . . . We tinker with your philosophy by direct manipulation of your cognitive experience. . . . It takes only a tiny group of engineers to create technology that can shape the entire future of human experience with incredible speed.

The internet came into wide use after Postman wrote his book. But it certainly resonates as a psychological assessment of the world in which we live today. Reality in the contemporary world is shaped more and more by what is on the screen. In 1999, the movie *The Matrix* heralded in cinematic form the arrival of Postman's technopoly.[27] The movie provided a discerning perspective of how the world had changed after the invention of the internet. Like the main protagonist of that movie, Neo, we now live "on" and "through" the computer screen, and our engagement with reality is largely shaped by that screen, whose technical name is the *matrix*, as the network of circuits that defines computer technology is called. But the same word also meant "womb" in Latin. The movie's transparent subtext was that, with the advent of cyberspace, new generations are born through two kinds of wombs—the biological and the technological.

It is little wonder that the interference into the 2016 American presidential campaign by foreign actors occurred through the matrix, where truth and mendacity are no longer differentiated, and where people's minds can be easily "engineered," as Lanier so frighteningly pointed out (above). The cyberattack on America was a con job, whereby the hackers entered websites to manipulate the content deceptively. It is no stretch to say that Facebook brought victory to Trump's doorstep. Ironically, when Facebook came into wide use, around 2005, it was heralded as bringing about a liberation from conformity and a channel for expressing one's opinions freely and of encouraging the sharing of scientific and philosophical discourse. But this view is fast becoming an anachronism. Counting the number of friends is seemingly more of a goal than discussing philosophical or aesthetic matters. The internet has become addictive for many, as they seek out other people's reactions to their daily updated lives. Its negative effects on rational thinking are subtle, and thus more dangerous.

Digital natives who have grown up in the world of the matrix may feel that the kind of "reality" that unfolds on the internet is the only option available to them. In the past, social relations, enduring cultural traditions, and stable patterns of work, life, and leisure assured people that stable patterns of meaning and experience united them in real space. The internet has shattered this assurance, impelling individuals to develop new strategies to manage the shocks of everyday life.

Matrix theory, as it can be called for the sake of convenience, might explain why the lies of someone like a Trump, a veritable huckster, are accepted so easily by so many. His ideas are spread virally through the internet where they take on validity without any merit. Spanish sociologist Manuel Castells has cogently argued that the digital revolution has brought about unprecedented changes in how people desire to gain control over their sense of identity.[28] It has led to a tension that he labels the "net-versus-the self." The former constitutes the organizational structures that have emerged on the internet, and the latter, people's attempts to establish their identities (religious, ethnic, sexual, territorial, or national) in the new digital environment. It is no coincidence that Trump has become a master user of Twitter, knowing full well that people immerse themselves in the digital environment on a daily basis. Access to people's mind, Trump knows, is through the screen.

Marshall McLuhan claimed that the media in which information is recorded and transmitted are decisive in shaping trends and in charting future progress, since they extend human faculties significantly, recalibrating the mind and, thus, leading to a rewiring of the brain.[29] So, by simply switching on our television sets, visiting websites, or using social media, we tend to feel connected to others in an abstract, rather than real, fashion. The environment in which we interact, create, and express ourselves is no longer real space, but an electronic space where daily interactions and communications are becoming more and more virtual. In this space the mind is exposed to subtle engineering, because it tends not to filter the information, but to simply take it in unreflectively. The unexpected and unpredictable rise of Donald Trump was bolstered by this state of affairs. The hacking of the election played on resentments brewing in America between conservative beliefs and the new liberal practices that characterized previous liberal administrations. The ads used in the hacking undoubtedly affected minds, since all of us have become accustomed to accepting the information on a screen at face value. When challenged that we might have been duped by it, causing cognitive dissonance, we tend to reject the relevant evidence. After the election, several TV news organizations interviewed Trump supporters, challenging them with the "fact" that they were influenced by the hacking. It comes as no surprise that most of them rejected the challenge, saying that the information was true. Others admitted that it may have been false, but that it still "told it like it is."

Virtually everything that Trump says or does is fake, starting with his "Make America Great Again" slogan, which is one word short of plagiarism, since it was Ronald Reagan who coined the slogan "Let's Make America Great Again."[30] But in a world governed by lies and doublespeak, people believe that Trump originated it, as he falsely claims. The meaning of MAGA is pure doublethink, as discussed, because it is impossible to pin it down. It can mean anything to anyone. For this reason it dispels the cognitive dissonance that might emerge when Trump's lies are exposed. It can easily be made to fit into any ideology or mythology.

MACHIAVELLIAN INTELLIGENCE

Before concluding this foray into the Art of the Lie, it is relevant to consider more closely the concept of Machiavellian Intelligence since it suggests that evolution was the source of mendacity in our species. Overall, it does not hold water, since lying is not found across all primate species, nor across all human cultures. But one tenet of the proponents of Machiavellian Intelligence is worth considering. They claim that lying is more of a male predisposition than it is a female one. Actually, there is relevant psychological evidence that Machiavellianism is, in fact, likely to be more frequent in men than in women, even though it can occur in anyone, even children. That is, everyone lies, no matter their gender, but not everyone is a master liar. The latter tends to be male, whatever the biological or cultural reason may be. This theory has been implicitly adopted in this book by using the pronoun "he" in reference to the master liar. And indeed the master liars of history were typically male. Certainly, that is what Machiavelli thought.

Was this a misogynistic interpretation of history? Were there not master liars of history who were women? Significantly, Machiavelli was widely condemned as a misogynist, although this might be somewhat incorrect, as Michelle Tolman Clarke has cogently argued:[31]

> [Machiavelli's] three central political works feature dozens of women who engage in efficacious and often praiseworthy political action. To appreciate fully the character and value of their activity and ultimately Machiavelli's views on women as potential political agents, one must first carefully attend to his conception of *animo*. Usually translated as "spiritedness," *animo* represents the natural assertiveness, energy, and resoluteness that forms the basis of *virtù* if properly disciplined—usually by a city's modes and orders. By examining the plight of women, however, Machiavelli turns to those persons who stand outside the city's political institutions and thus tend to exercise unbridled *animo*,

for better or for worse. In addition to revealing his deep preoccupation with political outsiders, Machiavelli's appreciation of the political problems associated with womanhood also discloses one of his most radical impieties—the denaturalization of gender norms, an impiety we are only beginning to appreciate today.

It is useful to revisit briefly the mythic story of Cassandra. She was the daughter of King Priam and Queen Hecuba of Troy. According to Homer, her beauty was so overwhelming that Apollo fell madly in love with her, giving her the power to predict the future. But she did not return his love, so Apollo punished her by ordering that no person would ever believe her prophecies. So, Cassandra warned the Trojans to give Helen back to the Greeks and to beware of the Trojan Horse. But they did not believe her, leading to disastrous consequences. She was praying at the altar of Athena when Troy fell, and Agamemnon took her to Mycenae as a slave. There she was murdered. Cassandra is clearly symbolic of the victimization of women, as Florence Nightingale argued in her book, *Suggestions for Thought to Searchers after Religious Truth*.[32] It became evident during the American presidential election campaign that a Cassandra syndrome is still in us, whereby people are conditioned to not believe women in matters of political importance.

The question becomes: Could there ever be a female Mussolini or Trump? If the answer is in the negative, then the search for an answer as to why men are expected to be the prince-liars may have to look at biology, as De Waal and others have done. The claim is in fact made by proponents of Machiavellian Intelligence that lying is how the male of the species adapts to the world. It is therefore a genetic, rather than cultural, trait. The North American biologist E. O. Wilson, too, sees a trait such as lying as a result of male biology being based in evolutionary fitness.[33] This raises a key question: Is the structure of the male brain better suited for the Machiavellian Art of the Lie than the female brain?[34] While such an evolutionary explanation is pure speculation, by comparing the behaviors of males and females in the domain of mendacity, the differences are not only noticeable but also highly suggestive.

Whatever the truth may be, it is certainly valid to say that there seems to be a double standard when it comes to the use of lying in politics. In this arena, women are typically evaluated differentially from males. Consider the election campaign between Trump and Clinton. Had she used the same kind of hyperbolic language and mendacity as Trump, she would have likely been evaluated much more harshly than Trump. There was an implicit expectation for her to be more "womanly," which probably was a factor in her defeat. The implications of gender politics are enormous and cannot be entertained in any detail here. This is not to say that women have not been proficient liars and deceivers—they can be as deceitful as men, and in some cases even more so. But the mendacity and deceit manifested by a Trump or a Mussolini is not expected of them. So, maybe it is more true to say that gendered views of mendacity are hardly biological but based in culture and historical traditions. If a woman is conniving, then she must be so in a subtle way. Again, one could claim that Clinton lost the election in part because of the false MAGA narrative, but in larger part because she was a woman who was expected by culture to "behave" like one. The double standard is alive and well in the twenty-first century.

There are many more critiques of the concept of Machiavellian Intelligence that cannot be covered here. Evidence exists that this type of wily ingenuity correlates with the anticipation of future events and decision making. Already in childhood, we learn that deceit, lies, and manipulation can lead to social success. Lying is thus a developed trait in a cultural context, making it possible to manage our own emotions and to recognize the emotions of others. This might explain why Trump is so effective at getting his message across—he anticipates the emotion of his listeners, identifying what role they play in different situations.

Moreover, one can never underestimate the human ability to be inventive and creative, in good and bad ways. In this sense, lying may be an "art" after all. In fact, having described Trump as a Machiavellian liar-prince, the question arises that, as a huckster and showman *à la* Barnum, he may simply be acting as he did on his Reality TV show, *The Apprentice.* This is more than a possibility, as James Piereson has perceptively written:[35]

> If Trump goes down in failure, it will more likely be due to a slowdown in the economy or to some misstep by the Federal Reserve Board, or to some international incident that he cannot handle, or in any case to events not related to his character. Why? Because Trump's character, far from something that is hardwired internally and beyond his conscious control, may be a mask that he changes to suit the circumstances or his interests. Is Trump perhaps, then, the ultimate Machiavellian, pretending to be a demagogue, a crude and tasteless public figure like many of our Hollywood celebrities, all for the purpose of achieving some large service on behalf of his country? That is also a possibility worth considering, in which case he would deserve to go down in history as one of the great actors of all time. In a strange way, Trump seems to know what he is doing, even if everyone else thinks he is unhinged or out of control. He also appears to be comfortable in his own skin, likewise a useful quality in a first-rate actor. After all, in a time when celebrity intersects closely with politics, it is possible to think that the Donald Trump we see on stage is not the real Donald Trump at all, but a public concoction made out both to satisfy and to confront the bizarre culture in which we live.

EPILOGUE

In Greek myth, the god Dolos ("Deception") was a mythic trope for mendacity, trickery, guile, cunning, craftiness, dissimulation, and all the other tactics that make up the Art of the Lie wrapped into one. He attempted to make a fake copy of the statue of Aletheia, the goddess of truth, so as to deceive people into believing that they were looking at the real statue.

Prometheus was taken by the similarity between the real and fake statues, catapulting Dolos to the title of master of deception—the first true mythic Artist of the Lie. The Greeks recognized the fact that liars are part of humanity, perhaps suggesting unconsciously that without them we would not know what truth is. This may be why, every once in a while, a Dolos appears in human life to drive home this principle of human life. Perhaps this is where to best insert an American huckster like Donald Trump. He is a larger-than-life figure who is admired, feared, and hated all at once. At the same time he engenders social self-analysis, allowing many people to think about truth, history, and the future.

The story of Dolos is an essay on Machiavellianism, and how it can be easily used by someone to cause people into seeing what is not there. The liar-prince is a master illusionist, as mentioned several times, using techniques that hide his true intents and which produce Orwellian "alternate realities" into which he thrusts his victims. The master liar is also a consummate actor or performer and a master of hyperbole. Hillary Clinton's campaign slogan of "Ready for Her" was static, specific, and exclusive, unlike the powerful ambiguity of the MAGA slogan. Clinton's slogan also centered on "her" as an individual, thus highlighting her specifically, rather than a cause, as did MAGA. This created great resentment among some voters, who saw her as the prototypical "elite" who was wrapped up in herself and in political correctness. Trump appeared to rise above this, becoming a kind of antihero who would smash to smithereens all the pseudo-norms of the Clinton and Obama regimes—a destroyer who alone could obliterate and even wipe off the face of the earth relativism, political correctness, and postmodernism. Among the first to catch on and be duped by this act were the evangelicals, as mentioned several times. They saw him as a King Cyrus who came to earth to set things right. Trump's take-no-prisoners approach emerged, in fact, at a time when dissatisfaction with political correctness and its sanitized language had started to frustrate many people. Simply put, Donald Trump's big-tent, nonexclusive, and linguistically belligerent, macho campaign helped to attract, energize, and convince voters that the time to "Make America Great Again," to "Build the Wall," and to "Drain the Swamp" was long overdue.

Trump's ascendancy to power was gained through verbal warfare, not through any coup d'état or military invasion. He gained it through falsehoods, bullshit, and a general carnivalesque hucksterism that transformed him both into an antiheroic figure and a performer all wrapped into one. His words are self-contained and self-referential, reminiscent of Humpty Dumpty's explanation of his language in Lewis Carroll's *Through the Looking Glass*:[36]

> *Humpty Dumpty*: "When I use a word, it means just what I choose it to mean—neither more nor less."
> *Alice*: "The question is whether you can make words mean so many different things."
> *Humpty Dumpty*: "The question is which is to be master—that's all."

Trump is a contemporary Humpty Dumpty, who knows how to get people to accept his version of reality, and to those who see through him, he responds with attacks or charges of "fake news," "witch hunts," or "hoaxes." But there is a cautionary note for Trump himself, as Carroll knew, and as a popular nineteenth-century nursery rhyme about Humpty Dumpty so aptly put it:

> Humpty Dumpty sat on a wall
> Humpty Dumpty had a great fall
> All the king's horses and all the king's men
> Could not put Humpty Dumpty together again.

Machiavelli's *The Prince* can be read both as a treatise on lying and a portrait of the human mind, in its darkest corners, but also an implicit Humpty Dumpty story—sooner or later the liar-prince will be unmasked and suffer a "great fall." The eighteenth-century Italian philosopher

Giambattista Vico, who saw history as moving in cycles,[37] developed a tripartite theory of cultural stages, or *corsi* (courses)—the divine, the heroic, and the human. Each stage manifested its own particular kind of customs, laws, language, and forms of consciousness. Vico did not, however, see this historical sequence as irreversible. So, he elaborated the idea of the *ricorso*, the return of an earlier age in the life of a culture. The course of humanity, according to Vico, goes from a divine form, through a heroic one, and finally to a rationalistic one. The divine stage generates myths; the heroic one, legends; and the rational one, factual history. Rationality, according to Vico, is humanity's greatest achievement. But unlike Cartesian philosophers, he did not see it as an innate given. He considered it to be a point of arrival that was achievable only in a social ambiance. Human beings do not inherit rationality from their biological legacy. Stripped of culture, which is a collective *memoria*, human beings would be forced to resort to their mythic imaginations to make sense of the world all over again.

So, it could well be that we are in a *ricorso*, whereby the relation of the parts (human subjectivities) to the whole (human collectivities) are evolving through unconscious historical forces, which play out in tandem with, but also separately from, technological and biological forces. In this framework, the *ricorso* may have been activated and a renewed humanistic consciousness is being retrieved. Eventually a Dolos culture will give way to an Aletheia culture. Mussolini and Hitler fell off the wall because people eventually came out of the darkness of their mythologies into the light of understanding. In his *Republic*, Plato portrayed humanity as imprisoned in a cave where it mistook shadows on the wall for reality.[38] Only the person with the imagination and courage to escape from the cave—the true philosopher—had the perspicacity to see the real world outside. By analogy, it can be said that the shadowy environment of the cave symbolizes the realm of mendacity. Trumpism exists in the cave, but, sooner or later, people will come out of it and "see the light," to use a cliché.

The metaphor of the cave leads to a historically inescapable conclusion—ultimately human beings will come down on the side of truth emerging from the cave into the light of truth, to reiterate the cliché. The concept of Machiavellian Intelligence is misguided in the end. Mendacity, in my view, is not a hard-wired trait. It is a result of human ingenuity that can be attenuated and even eliminated. The work on the theory of *autopoiesis* is of relevance here. The term was introduced by Maturana and Varela in their famous 1973 book, *Autopoiesis and Cognition*, where they claimed that an organism participates in its own evolution, since it has the ability to produce, or at least shape, its various biochemical agents and structures, thus ensuring their efficient and economical operation.[39] In the case of the human mind, autopoiesis seems to know no bounds. It is an acknowledgment of the infinite and flexible capacity of the human imagination to produce and reproduce knowledge and insights in its own creative way. Genetic factors alone do not completely define human beings. They tell us nothing about why humans create their meaningful experiences and pose the questions they do about life or why they might end up like Humpty Dumpty.

The lesson of all failed social and political experiments in "mind control," such as nazism, fascism, and communism, is that Machiavellianism eventually breaks down in front of honesty and truth. Despots cannot simply confine and imprison the human imagination for very long. The ancient Israelites used the word *hychma* to describe this aspect of human nature, defining it as the "science of the heart." That science will not let any lying scoundrel steal the heart of humanity. So too the Bible warns us in one of the Ten Commandments that "Thou shalt not bear false witness against thy neighbor," warning us that betrayal, mendacity, and all the other techniques of deceit must be overcome by integrity.

It is interesting to consider, in hindsight (and with foresight), that the ancient Greek civilization was founded by a woman, Athena, who came to earth to dispel the ugly machinations of men and set the world aright, in part by eliminating deceit and instilling *lógos* as intrinsic to social life.

Zeus entrusted Athena with his shield and his principal weapon, the thunderbolt. Her temple, the Parthenon, was in Athens (named after her). From there she gained enormous power over the world, becoming the goddess of cities and the arts, and, in later mythology, of wisdom. In a phrase, the ancient Athenian culture, with all its accomplishments, sprung from the wisdom of a goddess, who embodied truth, integrity, and wisdom. To use Lord Byron's marvelous poem, *Childe Harold's Pilgrimage* (canto 2, stanza 2), the modern world needs Athena again:[40]

> Ancient of days! august Athena! where,
> Where are thy men of might? thy grand in soul?
> Gone—glimmering through the dream of things that were.

The 2018 midterm elections in the United States could be interpreted as an "Athenian revolt," given the large number of women elected to the House of Representatives. Byron's plea is seemingly not falling on deaf ears today, over a century since he made it. The Czech writer Milan Kundera has also expressed a similar sentiment as follows:[41]

> Woman is the future of man. That means that the world which was once formed in man's image will now be transformed to the image of woman. The more technical and mechanical, cold and metallic it becomes, the more it will need the kind of warmth that only the woman can give it. If we want to save the world, we must adapt to the woman, let ourselves be led by the woman, let ourselves be penetrated by the *Ewigweiblich*, the eternally feminine!

The liar-prince takes away peace of mind, generates a negative karma, and pits people against each other, stirring hatred and fear, the darkest emotions that are found in the "cave" of the mind. By understanding what the Art of the Lie is all about and what it does, then we can unmask the liar and regain a sense of balance, much like the world described in one of Hans Christian Andersen's most significant cautionary fables, *The*

Emperor's New Clothes.[42] It is fitting to conclude with a paraphrase of the story, since it is more relevant today than it ever was. Two weavers promise the emperor that they will make him a new suit of clothes that they say, deceptively, is invisible to those who are incompetent and unfit for their positions. In reality, they make no clothes at all, making everyone *believe* that the clothes are invisible to them. When the emperor walks before his subjects, no one dares admit that he is nude for fear of reprobation or being labeled as incompetent. It took a child to cry out, at long last, "But he isn't wearing anything at all." The moral of the story is certainly valid in the current climate populated by clothes-less emperors around the world.

The truth must come out, and it is the only antidote to falsehoods; it is the only path to maintaining freedom of mind, leading us out of Plato's cave. Calling a consummate liar out is the only true counterattack we have, and must continue to employ, even if the truth may be upsetting. As Orwell so eloquently put it: "If liberty means anything at all it means the right to tell people what they do not want to hear."[43]

NOTES

PREFACE

1. Said to Richard Nixon (Tape of March 21, 1973).
2. Cited in Jane Mayer, "Donald Trump's Ghostwriter Tells All," *The New Yorker*, July 18, 2016. https://www.newyorker.com/magazine/2016/07/25/donald-trumps-ghostwriter-tells-all.
3. @tonyschwartz, 16 June 2015, 5:15 p.m.
4. Cited in David Brennan, "'Art of the Deal' Co-Author Says He Would Like to Rename the Book 'The Sociopath.'" *Newsweek* May 20, 2019. https://www.newsweek.com/art-deal-co author-rename-book-sociopath-1420701.
5. Donald Trump with Tony Schwartz, *The Art of the Deal* (New York: Ballantine Books, 1987), 58.
6. *Seinfeld*, season 6, episode 15, "The Beard," directed by Andy Ackerman, written by Larry David, Jerry Seinfeld, and Carol Leifer, aired February 9, 1995, on CBS.

CHAPTER 1

1. Peter Walcott, "Odysseus and the Art of Lying," *Ancient Society* 8 (1977): 1–19.
2. See Richard W. Byrne and Andrew Whiten, eds., *Machiavellian Intelligence: Social Expertise and the Evolution of Intellect in Monkeys, Apes, and Humans* (Oxford: Oxford University Press, 1988).
3. Robert Wright, *The Moral Animal: Why We Are the Way We Are* (New York: Vintage, 1995).
4. All quotations from the Bible are from the King James version.
5. Quoted in Jean Aitchison, *The Seeds of Speech: Language Origin and Evolution* (Cambridge: Cambridge University Press, 2000), 65.
6. Sun Tzu, *The Art of War* (N. Chelmsford, MA: Courier Corporation Reprint, 2002), 42.
7. Aristotle, *The Nicomachean Ethics* (Oxford: Oxford University Press, 2009).
8. Immanuel Kant, *Critique of Pure Reason*, trans. N. Kemp Smith (1781; repr., New York: St. Martin's, 2003).

9. See, for example, Sergey Gavrilets and Aaron Vose, "The Dynamics of Machiavellian Intelligence," *Proceedings of the National Academy of Sciences of the United States,* 103, no. 45 (November 7, 2006): 16823–28.

10. See Tamas Bereczkei, "Machiavellian Intelligence Hypothesis Revisited: What Evolved Cognitive and Social Skills May Underlie Human Manipulation," *Evolutionary Behavioral Sciences,* 12, no. 1 (January 2018): 32–51.

11. Richard Byrne, *The Thinking Ape: The Evolutionary Origins of Intelligence* (Oxford: Oxford University Press, 1995), 196.

12. Frank Nuessel, "Lying—A Semiotic Perspective," *Semiotics 2013*, Semiotic Society of America, 2013, 151–62.

13. This was conducted informally with three students of mine who are native speakers of these languages and through an internet search of terms. It is in no way a scientific statistical analysis; it was meant simply to get a sense of how different languages might converge in this area of vocabulary.

14. As mentioned, it is unclear who said this, and where it comes from.

15. Hannah Arendt, "Hannah Arendt: From an Interview," *New York Review of Books*, October 26, 1978, https://www.nybooks.com/articles/1978/10/26/hannah-arendt-from-an-interview/. Excerpted from a 1974 interview with Roger Errera.

16. Plato, *Phaedrus* [370 BCE], in *The Phaedrus, Lysis, and Protagoras of Plato*, trans. J. Wright (London: Macmillan and Co., 1925), 78.

17. Desiderius Erasmus, *Praise of Folly*, trans. Betty Radice (1509; London: Penguin Books, 1993), 71.

18. Niccolò Machiavelli, *The Prince* [1513], trans. W. K. Marriott, 1908, available online at Project Gutenberg, chap. 18, http://www.gutenberg.org/files/1232/1232-h/1232-h.htm.

19. Friedrich Nietzsche, *Human, All Too Human*, trans. Alecander Harvey (Chicago: Charles H. Kerr, 1908).

20. Machiavelli, *Prince.*

21. Ibid.

22. Ibid.

23. Antonio Nicaso and Marcel Danesi, *Made Men: Mafia Culture and the Power of Symbols, Rituals, and Myth* (Lanham: Rowman & Littlefield, 2013), 4.

24. Diego Gambetta, *The Sicilian Mafia: The Business of Private Protection* (Cambridge: Harvard University Press, 1993), 139.

25. Edward Sapir, "The Status of Linguistics as a Science," *Language* 5, no. 4 (December 1929): 209.

26. Paul Lunde, *Organized Crime: An Inside Guide to the World's Most Successful Industry* (London: Dorling Kindersley, 2004), 54.

27. Antonino Cutrera, *La mafia e i mafiosi* (Palermo: Reber, 1900), 2 (translated by the author).

28. Tzu, *Art of War*, 40.

29. The phrase was also used during the French Revolution, and became a slogan since the inception of the Soviet Union, starting with Lenin in a decree of November 28, 1917. See Nicholas Werth et al., *The Book of Communism: Crimes, Terror, Repression*. Cambridge, MA: Harvard University Press, 1999.

30. Oscar Wilde, "The Decay of Lying," in *Complete Works*, ed. Josephine M. Guy, vol. 4, *Criticism: Historical Criticism, Intentions, the Soul of Man* (Oxford: Oxford University Press, 2007), 102.

31. Ibid.

32. Ibid., 95.

33. Aldert Vrij, Pär Anders Granhag, and Stephen Porter, "Pitfalls and Opportunities in Nonverbal and Verbal Lie Detection," *Psychological Science in the Public Interest* 11, no. 3 (December 2010): 89–121.

34. Samuel Butler, *Notebooks* (1919; repr., New York: E. P. Dutton, 1951), 220.

35. Hadley Cantril, *The Invasion from Mars: A Study in the Psychology of Panic* (Princeton, NJ: Princeton University Press, 1940).

36. Mikhail Bakhtin, *The Dialogic Imagination: Four Essays* (Austin: University of Texas Press, 1981); Mikhail Bakhtin, *Rabelais and His World* (Bloomington: Indiana University Press, 1984).

37. Donald Trump, "Remarks by President Trump on Antiquities Act Designations," Salt Lake City, UT, December 4, 2017, https://www.whitehouse.gov/briefings-statements/remarks-president-trump-antiquities-act-designations/.

38. Donald Trump, "Remarks by President Trump at Signing of Executive Order on the Antiquities Act," Washington, DC, April 26, 2017, https://www.whitehouse.gov/briefings-statements/remarks-president-trump-signing-executive-order-antiquities-act/.

39. Donald J. Trump (@realDonaldTrump), Twitter, August 6, 2018, 4:35 p.m., https://twitter.com/realdonaldtrump/status/1026587142989008897?lang=en.

40. Claude Lévi-Strauss, *La pensée sauvage* (Paris: Plon, 1962).

41. Wilson Bryan Key, *The Age of Manipulation: The Con in Confidence, the Sin in Sincere* (New York: Henry Holt, 1989), 13.

42. Oscar Wilde, "Aristotle at Afternoon Tea," *Pall Mall Gazette* (London), December 16, 1887, 3.

43. Amanda Carpenter, *Gaslighting America: Why We Love It When Trump Lies to Us* (New York: HarperCollins, 2018).

44. Ruth Ben-Ghiat, "An American Authoritarian," *Atlantic*, August 10, 2016, https://www.theatlantic.com/politics/archive/2016/08/american-authoritarianism-under-donald-trump/495263/.

45. "Full Text: Donald Trump 2016 RNC Draft Speech Transcript," *Politico*, July 21, 2016, https://www.politico.com/story/2016/07/full-transcript-donald-trump-nomination-acceptance-speech-at-rnc-225974.

46. B. Joey Basamanowicz, *Believe Me: 21 Lies Told by Donald Trump and What They Reveal about His Vision for America* (Scotts Valley, CA: CreateSpace, 2016).

47. James W. Pennebaker, *The Secret Life of Pronouns: What Our Words Say about Us* (London: Bloomsbury, 2011).

48. Donald J. Trump (@realDonaldTrump), Twitter, May 8, 2013, 6:37 p.m., https://twitter.com/realdonaldtrump/status/332308211321425920?lang=en.

49. Donald J. Trump (@realDonaldTrump), Twitter, May 13, 2015, 5:23 a.m., https://twitter.com/realdonaldtrump/status/598463630288560128.

50. Donald J. Trump (@realDonaldTrump), Twitter, July 21, 2014, 1:50 p.m., https://twitter.com/realdonaldtrump/status/491324429184823296?lang=en.

51. Donald J. Trump (@realDonaldTrump), Twitter, January 23, 2016, 5:57 a.m., https://twitter.com/realdonaldtrump/status/690896236765806592?lang=en.

52. Donald J. Trump (@realDonaldTrump), Twitter, March 23, 2016, 12:54 p.m., https://twitter.com/realdonaldtrump/status/712729376207540226?lang=en.

53. Donald J. Trump (@realDonaldTrump), Twitter, January 6, 2018, 7:27 a.m., https://twitter.com/realdonaldtrump/status/949618475877765120?lang=en.

54. Donald J. Trump, interview by Chuck Todd, *Meet the Press*, NBC, February 28, 2016, transcript, https://www.nbcnews.com/meet-the-press/meet-press-february-28-2016-n527506.

55. Donald J. Trump (@realDonaldTrump), December 22, 2015, 5:19 p.m.

56. Donald J. Trump (@realDonaldTrump), February 28, 2016, 7:28 a.m.

57. Donald J. Trump (@realDonaldTrump), April 17, 2016, 9:41 a.m.

58. Presidential announcement speech, June 16, 2015.

59. Donald J. Trump (@realDonaldTrump), April 6, 2015, 11:22 p.m.

60. Madeleine Albright, *Fascism: A Warning* (New York: HarperCollins, 2018).

61. John Kelly, "What's With All Trump's Talk about 'Draining the Swamp?'" *Slate*, October 26, 2016, https://slate.com/human-interest/2016/10/why-do-trump-and-his-supports-keep-talking-about-draining-the-swamp.html.

62. Marshall McLuhan, *The Gutenberg Galaxy* (Toronto: University of Toronto Press, 1962).

63. Karl Marx, *The Economic and Philosophical Manuscripts of 1844*, trans. Martin Milligan (New York: International, 1964).

64. Émile Durkheim, *The Elementary Forms of Religious Life*, trans. Joseph Ward Swain (New York: Macmillan, 1912).

65. Harriet Sherwood, "'Toxic Christianity': The Evangelicals Creating Champions for Trump," *Guardian*, October 21, 2018, https://www.theguardian.com/us-news/2018/oct/21/evangelical-christians-trump-liberty-university-jerry-falwell.

66. Niccolò Machiavelli, *The Art of War* [1521] (Scotts Valley, CA: CreateSpace, 2010).

67. George Orwell, *Orwell on Truth*, 75.

68. "Bernstein: We're in a Cold Civil War," CNN, September 25, 2018, quoted in Ben Yagoda, "Are We in a 'Cold Civil War'?" *Chronicle of Higher Education*, October 7, 2018, https://www.chronicle.com/blogs/linguafranca/2018/10/07/are-we-in-a-cold-civil-war/.

69. Yagoda, "Are We in a 'Cold Civil War.'"

70. David Frum, *Trumpocracy: The Corruption of the American Republic* (New York: HarperCollins, 2018), 108.

71. Marcel Proust, *Remembrance of Things Past*, vol. 5 (1913; repr., Harmondsworth, UK: Penguin, 1983), p. 515.

72. Walter Lippmann, *Public Opinion* (New York: Macmillan, 1922).

73. Harold Laswell, *Propaganda Technique in World War I* (Cambridge, MA: MIT, 1971).

74. Ralph Waldo Emerson, "Prudence," in *Essays* (New York: Charles E. Merrill, 1907), 256, available online at Project Gutenberg, https://www.gutenberg.org/files/16643/16643-h/16643-h.htm.

75. Sherry Turkle, *Reclaiming Conversation: The Power of Talk in a Digital Age* (New York: Penguin, 2015), 297.

76. Aristotle, *Rhetoric*, in *The Works of Aristotle*, ed. W. D. Ross, vol. 11 (Oxford: Clarendon Press, 1952), p. 62.

77. Norman Mailer, *Advertisements for Myself* (1959; repr., Cambridge, MA: Harvard University Press, 1992), 23.

CHAPTER 2

1. George Orwell, *1984* (London: Secker and Warburg, 1949).

2. Ibid., chap. 3.

3. Ralph Benko, "The Left, Not Kellyanne Conway, Invented 'Alternative Facts,'" *Forbes*, February 11, 2017, https://www.forbes.com/sites/ralphbenko/2017/02/11/the-left-not-kellyanne-conway-invented-alternative-facts/#d7188c0658c6.

4. Edward S. Herman, *Beyond Hypocrisy: Decoding the News in an Age of Propaganda Including a Doublespeak Dictionary for the 1990s* (Montreal: Black Rose, 1992), 3.

5. Edward Herman and Noam Chomsky, *Manufacturing Consent: The Political Economy of the Mass Media* (New York: Pantheon, 1988).

6. Cited in Alfred B. Evans, *Soviet Marxism-Leninism: The Decline of an Ideology* (Westport, CT: Greenwood, 1993), 39.

7. Herman, *Beyond Hypocrisy*.

8. Orwell, *1984*, 32.

9. Donald J. Trump, from a speech delivered at a Veterans of Foreign Wars Convention in Kansas City, Missouri, https://www.whitehouse.gov/briefings-statements/remarks-president-trump-veterans-foreign-wars-united-states-national-convention-kansas-city-mo/.

10. Ibid., book 1, chap. 7.

11. Donald J. Trump (@realDonaldTrump), Twitter, December 28, 2017, 4:01 p.m., https://twitter.com/realdonaldtrump/status/946531657229701120?lang=en.

12. See Benito Mussolini, *Mussolini as Revealed in His Political Speeches, November 1914–August 1923*, trans. Barone Bernardo Quaranta (Charleston, SC: Nabu, 2010); Benito Mussolini, *Selected Speeches of Benito Mussolini* (Amazon Digital Services, 2018).

13. Arthur Asa Berger, *Ads, Fads, and Consumer Culture: Advertising's Impact on American Character and Society* (Lanham: Rowman & Littlefield, 2000), 131.

14. Orwell, *1984*, book 1, chap. 8.

15. Chaim Shinar, "Conspiracy Narratives in Russian Politics: From Stalin to Putin," *European Review* 26 (2018): 648–60.

16. See Antonio Nicaso and Marcel Danesi, *Made Men: Mafia Culture and the Power of Symbols, Rituals, and Myth* (Lanham: Rowman & Littlefield, 2013).

17. See Salvatore di Piazza, Francesca Piazza, and Mauro Serra, "The Need for More Rhetoric in the Public Sphere: A Challenging Thesis about Post-Truth," *Versus* 127 (2018): 225–42.

18. See María José Martín-Velasco and María José García Blanco, *Greek Philosophy and Mystery Cults* (Cambridge: Cambridge Scholars, 2016).

19. George Orwell, *The Collected Essays*, ed. Sonia Orwell and Ian Angus, vol. 3 (London: Secker and Warburg, 1968), 6.

20. Michael Barkun, *A Culture of Conspiracy: Apocalyptic Visions in Contemporary America* (Berkeley: University of California Press, 2003), 3–4.

21. Ibid.

22. J. P. Linstroth, "Myths on Race and Invasion of the 'Caravan Horde,'" *Counterpunch*, November 9, 2018, https://www.counterpunch.org/2018/11/09/myths-on-race-and-invasion-of-the-caravan-horde/.

23. Richard Dawkins, *The Selfish Gene* (Oxford: Oxford University Press, 1976).

24. Marcel Danesi, *Memes and the Future of Popular Culture* (Leiden: Brill Academic, 2019).

25. Robert Paxton, *The Anatomy of Fascism* (New York: Vintage, 2004).

26. Jean de La Bruyère, *Characters* (New York: Scribner & Welford, 1885), 13.

27. Aldous Huxley, *Beyond the Mexique Bay* (London: Paladin, 1934), 12.

28. Emile Durkheim, *The Elementary Forms of Religious Life* (New York: Collier, 1912).

29. Donald J. Trump, uttered during the final televised presidential debate, October 19, 2016.

30. Adolf Hitler, *Mein Kampf* (Munich: Franz Eher Nachfolger, 1925), chap. 6.

31. Francesco Magiapane, "The Discourse of Fake News in Italy," *Versus* 127 (2018): 298.

32. Mussolini, quoted in *Selected Speeches of Benito Mussolini*.

33. Ibid.

34. CNN interview of a group of white evangelical women, https://www.cnn.com/videos/politics/2018/01/28/women-trump-supporters-reaction-stormy-daniels-kaye-dnt-nr.cnn.

35. Hesiod, *Theogony* (700 BCE; repr., Scotts Valley, CA: CreateSpace, May 12, 2017).

36. Donald J. Trump, uttered during a presidential campaign speech at Sioux City, Iowa, January 23, 2016.

37. Frederik H. Lund, "The Psychology of Belief," *Journal of Abnormal and Social Psychology* 20 (1925): 63. Perhaps the first ever psychological study of belief was by William James, "The Psychology of Belief," *Mind* 14 (1889): 321–53.

38. Charles S. Peirce, *Illustrations of the Logic of Science* (1877–1878; repr., Chicago: Open Court, 2014).

39. Martin Luther King Jr., Speech in Montgomery, Alabama, March 25, 1965, as transcribed from a tape recording.

40. Quoted in Steven R. Weiman, ed., *Daniel Patrick Moynihan: A Portrait in Letter of an American Visionary* (New York: PublicAffairs, 2010).

41. See Kay M. Porterfield, *Straight Talk about Cults* (New York: Facts on File, 1997).

42. Frank Nuessel, "Lying: A Semiotic Perspective," *Semiotics* (Semiotic Society of America, 2013): 151–62.

43. Orwell, *1984*, 37–38.

44. Cited in F. Scott Fitzgerald, "The Crack-Up," *Esquire Magazine*, February 1936.

CHAPTER 3

1. The idea of self-construction was explored in detail for the first time by psychologist Carl Rogers, "A Theory of Therapy, Personality and Interpersonal Relationships as Developed in the Client-Centered Framework," in *Psychology: A Study of a Science*, vol. 3, *Formulations of the Person and the Social Context*, ed. S. Koch (New York: McGraw Hill, 1959).

2. F. Max Müller, *Biographies of Words and the Home of the Aryas* (1888; repr., Whitefish, MT: Kessinger, 2004), 120.

3. Benito Mussolini, speech in Bologna, April 1921.

4. Antonio Nicaso and Marcel Danesi, *Made Men: Mafia Culture and the Power of Symbols, Rituals, and Myth* (Lanham: Rowman & Littlefield, 2013).

5. Ibid.

6. Ibid.

7. Paul Lunde, *Organized Crime: An Inside Guide to the World's Most Successful Industry* (London: Dorling Kindersley, 2004), 55.

8. Luigi Natoli, *Coriolano della Floresta* (1720; repr., Milano: Sellerio, 2017).

9. See Umberto Santino, *La cosa e il nome* (Soveria Mannelli: Rubbettino, 2000), 119–28.

10. John Lawrence Reynolds, *Shadow People: Inside History's Most Notorious Secret Societies* (Toronto: Key Porter, 2006), 177–78.

11. Claude Lévi-Strauss, *The Raw and the Cooked* (London: Cape, 1964), 23.

12. *The Birth of a Nation,* directed by D. W. Griffith, released by David W. Griffith Corp., on February 8, 1915; Thomas Dixon Jr. *The Clansman: A Historical Romance of the Ku Klux Klan* (New York: Grosset & Dunlap, 1905).

13. *Intolerance*, directed by D. W. Griffith, released by Triangle Distributing Corporation on September 5, 1916.

14. W. T. Anderson, *Reality Isn't What It Used to Be* (San Francisco: HarperCollins, 1992).

15. Marcel Proust, "The Fugitive," in *Remembrance of Things Past*, vol. 3 (1925; repr., New York: Random House, 1981).

16. Anderson, *Reality Isn't What It Used to Be*, 126–27.

17. Jean Baudrillard, *Simulations* (New York: Semiotexte, 1983).

18. *Mississippi Burning*, directed by Alan Parker, distributed by Orion Pictures Corporation on December 2, 1988.

19. *Ghosts of Mississippi*, directed by Rob Reiner, produced by Castle Rock Entertainment, distributed by Columbia Pictures on December 20, 1996.

20. *A Time to Kill*, directed by Joel Schumacher, produced by Regency Enterprises, distributed by Warner Brothers on July 24, 1996.

21. *BlacKkKlansman*, directed by Spike Lee, distributed by Focus Features on August 10, 2018.

22. Christian Broadcasting Network interview with Sarah Sanders, January 30, 2018, https://www1.cbn.com/cbnnews/politics/2019/january/exclusive-white-house-press-secretary-sarah-sanders-god-wanted-donald-trump-to-become-president.

23. Cited in George Orwell, *Orwell on Truth* (London: Harvill Secker, 2017), 173.

24. Abraham Lincoln, Proclamation of Thanksgiving, cited in Roy B. Basler et al., *Collected Works of Abraham Lincoln*, The Abraham Lincoln Foundation, 1953, http://www.abrahamlincolnonline.org/lincoln/speeches/thanks.htm.

25. Max Weber, *The Protestant Ethic and the Spirit of Capitalism* (New York: Scribner's, 1905).

26. Arthur Asa Berger, *Shop 'til You Drop: Consumer Behavior and American Culture* (Lanham: Rowman & Littlefield, 2005), 6.

27. Ibid., 7.

28. Niccolò Machiavelli, *The Prince*, trans. W. K. Marriott (1513), chap. 18, available online at Project Gutenberg, http://www.gutenberg.org/files/1232/1232-h/1232-h.htm.

29. Lao Tzu, *Tao Te Ching*, trans. Witter Bynner, June 15, 1944, https://www.aoi.uzh.ch/dam/jcr:ffffffff-c059-cfbc-0000-000033ea7e9a/BynnerLao.pdf.

30. Machiavelli, *The Prince*, chap. XVIII.

31. Frank Nuessel, "Deception and Its Manifestations," *Semiotics 2015: Virtual Identities* (Semiotic Society of America [2016]): 179.

32. Michel Foucault, *The History of Sexuality* (London: Allen Lane, 1976).

33. Desmond Morris, *The Human Zoo* (London: Cape, 1969).

34. Carl Sagan and Ann Druyan, *Shadows of Forgotten Ancestors: A Search for Who We Are* (New York: Random House, 1992), 415.

35. Benito Mussolini, cited in Mark Neocleous, *Fascism* (Minneapolis: University of Minnesota Press, 1997), 35.

36. See Meg Wagner et al., "Suspect Arrested after Explosive Devices Sent to Trump Critics and CNN," November 5, 2018, https://www.cnn.com/politics/live-news/clintons-obama-suspicious-packages/index.html.

37. See, respectively: "Trump Hails Body Slamming Congressman Greg Gianforte in Montana," BBC News, October 19, 2018, https://www.bbc.com/news/world-us-canada-45913921; Andrew Restuccia, "Trump Stays Silent on Media-Hating Coast Guard Officer," *Politico*, February 21, 2019, https://www.politico.com/story/2019/02/21/trump-coast-guard-officer-1179749.

38. George Orwell, *Orwell on Truth* (London: Harvill Secker, 2017), 11.

39. Machiavelli, *The Prince*, chap. XVIII.

40. Ibid.

41. Proni, *Talking to Oneself*, 3.

42. Arthur Asa Berger, *Three Tropes on Trump: Marxism, Semiotics and Psychoanalysis* (self-published, 2019), 48.

43. B. Joey Basamanowicz and Katie Poorman, *Believe Me: 21 Lies Told by Donald Trump and What They Reveal about His Vision for America* (Scotts Valley, CA: CreateSpace, September 2016).

44. See, "Clip of Presidential Candidate Donald Trump Rally in Mount Pleasant, South Carolina," Donald Trump, December 7, 2015, C-SPAN, video, 0:31, https://www.c-span.org/video/?c4737466/trumps-muslim-ban.

45. Cliff Sims, *Team of Vipers* (New York: Thomas Dunne, 2019).

46. William Cummings, "Conspiracy Theories: Here's What Drives People to Them, No Matter How Wacky," *USA Today*, January 15, 2018, https://www.usatoday.com/story/news/nation/2017/12/23/conspiracy-theory-psychology/815121001/.

47. Antonio R. Damasio, *Descartes' Error: Emotion, Reason, and the Human Brain* (New York: G. P. Putnam's, 1994).

48. Cummings, "Conspiracy Theories."

CHAPTER 4

1. Cited in Norman Mailer, *Advertisements for Myself* (1959; repr., Cambridge, MA: Harvard University Press reprint, 1992).

2. Samanth Subramanian, "Inside the Macedonian Fake-News Complex," *Wired*, February 15, 2017, https://www.wired.com/2017/02/veles-macedonia-fake-news/.

3. Piero Polidoro, "Post-Truth and Fake News: Preliminary Considerations," *Versus* 127 (2018): 189–206.

4. Kevin Young, "Moon Shot: Race, a Hoax, and the Birth of Fake News," *New Yorker*, October 21, 2017, https://www.newyorker.com/books/page-turner/moon-shot-race-a-hoax-and-the-birth-of-fake-news.

5. Reported by Matt Kwong, "These Trump Voters Support the U.S. President's Comments on Russia—and His Walkback, Too," CBC News, July 18, 2018, https://www.cbc.ca/news/world/trump-putin-us-maryland-essex-dundalk-edgemere-1.4751215.

6. Brooke Donald, "Stanford Researchers Find Students Have Trouble Judging the Credibility of Information Online," Stanford Graduate School of Information, November 22, 2016, https://ed.stanford.edu/news/stanford-researchers-find-students-have-trouble-judging-credibility-information-online.

7. Edward Herman and Noam Chomsky, *Manufacturing Consent: The Political Economy of the Mass Media* (New York: Pantheon, 1988).

8. Richard Wooley, "Donald Trump, Alex Jones and the Illusion of Knowledge," CNN, August 6, 2018, https://www.cnn.com/2017/07/15/opinions/trump-alex-jones-world-problem-opinion-wooley/index.html.

9. Gary Lachman, *Turn Off Your Mind: The Mystic Sixties and the Dark Side of the Age of Aquarius* (New York: Disinformation, 2001).

10. Ibid., 396–97.

11. Donald J. Trump (@realDonaldTrump), Twitter, June 18, 2018, 7:57 a.m., https://twitter.com/realdonaldtrump/status/1008725438972211200?lang=en.

12. Donald J. Trump (@realDonaldTrump), Twitter, July 3, 2018, 6:16 a.m., https://twitter.com/realdonaldtrump/status/1014105549624037377?lang=en.

13. Donald J. Trump (@realDonaldTrump), Twitter, June 25, 2018, 5:36 a.m., https://twitter.com/realdonaldtrump/status/1011226622324887556?lang=en.

14. Donald J. Trump (@realDonaldTrump), Twitter, June 20, 2018, 7:25 a.m., https://twitter.com/realdonaldtrump/status/1009411866475532288?lang=en.

15. Donald J. Trump (@realDonaldTrump), Twitter, June 4, 2018, 3:41 p.m., https://twitter.com/realdonaldtrump/status/1003738642903420928?lang=en.

16. Donald J. Trump (@realDonaldTrump), Twitter, June 13, 2018, 8:30 a.m., https://twitter.com/realdonaldtrump/status/1006891643985854464?lang=en.

17. Donald J. Trump (@realDonaldTrump), Twitter, May 18, 2018, 8:50 a.m.

18. Donald J. Trump (@realDonaldTrump), Twitter, May 4, 2018, 5:45 a.m., https://twitter.com/realdonaldtrump/status/992354530510721025?lang=en.

19. Donald J. Trump (@realDonaldTrump), Twitter, April 3, 2018, 5:34 a.m., https://twitter.com/realdonaldtrump/status/981117684489379840?lang=en.

20. e. e. cummings, cited in *Vanity Fair*, December 1926.

21. Hadley Cantril, *The Invasion from Mars: A Study in the Psychology of Panic* (Edison, NJ: Transaction, 1940).

22. Mark Dice, *The True Story of Fake News: How Mainstream Media Manipulates Millions* (San Diego: Resistance Manifesto, 2017).

23. For a comprehensive account of Stalin's use of disinformation and forgeries, see Mikhail Agursky, "Soviet Disinformation and Forgeries," *International Journal of World Peace*, 6 (1989): 13–30.

24. Benito Mussolini, quoted in *Selected Speeches of Benito Mussolini* (Amazon Digital Services, 2018).

25. Ibid.

26. Ibid.

27. Donald J. Trump, speech delivered at a North Carolina rally, August 18, 2016. Transcript: http://edition.cnn.com/TRANSCRIPTS/1608/18/cnnt.01.html.

28. Cited in Clive Irving, "Trump's War on the Press Follows the Mussolini and Hitler Playbook," *Daily Beast*, January 12, 2018, https://www.thedailybeast.com/trumps-war-on-the-press-follows-the-mussolini-and-hitler-playbook.

29. Ibid.

30. For a report of Conway's assertion, see Gideon Resnick, "Kellyanne Conway Cites Fake Bowling Green Massacre," *The Daily Beast*, February 2, 2017, https://www.thedailybeast.com/kellyanne-conway-cites-fake-bowling-green-massacre?source=twitter&via=desktop.

31. Niccolò Machiavelli, *On Conspiracies* (1513; repr., Harmondsworth: Penguin, 2010).

32. Alessandro Campi, *Machiavelli and Political Conspiracies* (New York: Routledge, 2018).

33. Karl Popper, *The Open Society and Its Enemies* (London: Routledge, 1945).

34. For a comprehensive analysis of the influence of Father Coughlin on the spread of the fake news syndrome, see Jack Kay, George W. Ziegelmueller, and Kevin M. Minch, "From Coughlin to Contemporary Talk Radio: Fallacies & Propaganda in American Populist Radio," *Journal of Radio Studies* 5 (1998): 9–21.

35. Chip Heath and Dan Heath, *Made to Stick: Why Some Ideas Survive and Others Die* (New York: Random House, 2006).

36. Joseph Heller, *Catch-22* (New York: Simon & Schuster, 1955).

37. Geoff Nunberg, "Why the Term 'Deep State' Speaks to Conspiracy Theorists," NPR, August 9, 2018, https://www.npr.org/2018/08/09/633019635/opinion-why-the-term-deep-state-speaks-to-conspiracy-theorists.

38. Sander van der Linden, "The Surprising Power of Conspiracy Theories," *Psychology Today*, August 24, 2015, https://www.psychologytoday.com/ca/blog/socially-relevant/201508/the-surprising-power-conspiracy-theories.

39. Henry Giroux, "Challenging Trump's Language of Fascism," *Truthout*, January 9, 2016, https://truthout.org/articles/challenging-trumps-language-of-fascism/.

40. *Oxford English Dictionary*, https://languages.oup.com/word-of-the-year/hub.

41. Lewis Carroll, *Through the Looking-Glass and What Alice Found There* (London: Macmillan, 1871), cited in chap. 5.

42. Jean Baudrillard, *Simulations* (New York: Semiotexte, 1983).

43. Susan Greenfield, *Mind Change* (New York: Random House, 2015), 241.

44. See, for example, Marshall McLuhan, *Understanding Media: The Extensions of Man* (Cambridge, MA: MIT Press).

45. Noam Chomsky, *Media Control: The Spectacular Achievements Propaganda* (New York: Seven Stories Press, 2002), 16.

46. Wooley, "Donald Trump, Alex Jones."

CHAPTER 5

1. Desiderius Erasmus, from *Adages*, trans. and annotated by Denis L. Drysdall, ed. John N. Grant (Toronto: University of Toronto Press, 2005), chap. 4.

2. Amanda Carpenter, *Gaslighting America: Why We Love It When Trump Lies to Us* (New York: HarperCollins, 2018).

3. Bryant Welch, *State of Confusion: Political Manipulation and the Assault on the American Mind* (New York: Macmillan, 2008), 23.

4. *Gaslight,* directed by George Cukor, based on the play *Gas Light* (1938) by Thomas Hamilton, produced by Metro-Goldwyn-Mayer, distributed by Loew's, Inc., May 4, 1944.

5. Carpenter, *Gaslighting America.*

6. Karen Grigsby Bates, "'Rapists,' 'Huts': Trump's Racist Dog Whistles Aren't New," NPR, *Codeswitch*, January 13, 2018, https://www.npr.org/sections/codeswitch/2018/01/13/577674607/rapists-huts-shitholes-trumps-racist-dog-whistles-arent-new.

7. Trump reportedly uttered the slur during a cabinet meeting with lawmakers on immigration (January 11, 2018), as reported by Avery Anapol, *The Hill*, https://thehill.com/home news/administration/368845-trump-defended-shithole-countries-remark-in-private-report.

8. Donald J. Trump (@realDonaldTrump), Twitter, January 12, 2018, 5:48 a.m., https://twitter.com/realdonaldtrump/status/951813216291708928?lang=en.

9. William Safire, *Safire's Political Dictionary* (New York: Oxford University Press, 2008), 190.

10. Amanda Lohrey, *Voting for Jesus: Christianity and Politics in Australia* (Melbourne: Black, 2006).

11. Niccolò Machiavelli, *Discourses on the First Decade of Titus Livius*, trans. Ninian Hill Thomson (London: Kegan Paul, Trench, 1883), chap. 46; available online at Project Gutenberg: http://www.gutenberg.org/cache/epub/10827/pg10827-images.html.

12. Cited in Eric Jabbari, *Pierre Laroque and the Welfare State in Postwar France* (Oxford: Oxford University Press, 2012), 46.

13. Benito Mussolini, quoted in *Selected Speeches of Benito Mussolini* (Amazon Digital Services, 2018).

14. Donald J. Trump, uttered at a rally in Houston on October 22, 2018, reported by John Walsh, "Trump Declares Himself a 'Nationalist' While Stumping for Ted Cruz," *Business Insider*, October 23, 2018, https://www.businessinsider.in/trump-declares-himself-a-nationalist-while-stumping-for-ted-cruz/articleshow/66327534.cms.

15. Bobby Azarian, "Trump Is Gaslighting America Again—Here's How to Fight It," *Psychology Today*, August 31, 2018, https://www.psychologytoday.com/ca/blog/mind-in-the-machine/201808/trump-is-gaslighting-america-again-here-s-how-fight-it.

16. Adolf Hitler, *Mein Kampf* (Munich: Franz Eher Nachfolger, 1925).

17. Walter C. Langer, *A Psychological Analysis of Adolph Hitler: His Life and Legend* (1943; repr., Scotts Valley, CA: CreateSpace, 2012).

18. See, for example, George Lakoff, *Women, Fire and Dangerous Things: What Categories Reveal about the Mind* (Chicago: University of Chicago Press, 1987); and Mark Johnson, *The Body in the Mind: The Bodily Basis of Meaning, Imagination and Reason* (Chicago: University of Chicago Press, 1987).

19. Donald J. Trump (@realDonaldTrump), Twitter, September 6, 2018, 4:19 a.m., https://twitter.com/realdonaldtrump/status/1037661562897682432?lang=en.

20. I. A. Richards, *The Philosophy of Rhetoric* (Oxford: Oxford University Press, 1936).

21. Richards, *The Philosophy of Rhetoric.*

22. Howard R. Pollio, Jack M. Barlow, Harold J. Fine, and Marilyn R. Pollio, *Psychology and the Poetics of Growth: Figurative Language in Psychology, Psychotherapy, and Education* (Hillsdale, NJ: Lawrence Erlbaum, 1977).

23. George Lakoff and Mark Johnson, *Metaphors We Live By* (Chicago: University of Chicago Press, 1980). Since the publication of this book, there has been a veritable upsurge in the study of figurative language within the cognitive sciences.

24. See, Robert Reich, "Trump Is Cornered, with Violence on His Mind. We Must Be on Red Alert," *The Guardian*, March 16, 2019, https://www.theguardian.com/us-news/commentisfree/2019/mar/16/donald-trump-breitbart-interview-white-supremacy.

25. Donald J. Trump (@realDonaldTrump), Twitter, January 18, 2018, 5:16 a.m., https://twitter.com/realdonaldtrump/status/953979393180950528?lang=en.

26. See Ryan Bort, "Lindsey Graham Said the Quiet Part Out Loud About the Border Wall," *Rolling Stone*, December 31, 2018, https://www.rollingstone.com/politics/politics-news/lindsey-graham-donald-trump-border-wall-773800/.

27. Rudyard Kipling, *Just So Stories for Little Children* (London: Macmillan Publishers, 1902).

28. See Aric Jenkins, "Donald Trump Called for NFL Players to Be Fired for National Anthem Kneeling—And They Responded," *Fortune.com*, September 23, 2017, https://fortune.com/2017/09/23/donald-trump-nfl-players-anthem-response/.

29. Cardinal de Richelieu, *Testament Politique*, "Maxims" (Amsterdam: Henry Desbordes, 1688).

30. Hannah Arendt, *The Origins of Totalitarianism* (New York: Harcourt, Brace, and Company, 1976), 350.

31. Donald J. Trump, Interview on CNN's "State of the Union"; see Timothy Cama, "Trump: 'We Certainly Do Have a Problem' with Some Muslims," *The Hill*, September 20, 2015,

https://thehill.com/blogs/blog-briefing-room/254307-trump-we-certainly-do-have-a-problem-with-some-muslims.

32. Niccolò Machiavelli, *The Prince*, trans. by W. K. Marriott (originally 1513), chap. 18; available online at Project Gutenberg: http://www.gutenberg.org/files/1232/1232-h/1232-h.htm.

33. See CNBC report, "Trump: 'Some Form of Punishment' Needed for Abortion," March 30, 2016, https://www.cnbc.com/video/2016/03/30/trump-some-form-of-punishment-needed-for-abortion.html.

34. Donald J. Trump, speech delivered at National Prayer Breakfast, 2019, see https://archive.org/details/CSPAN_20190208_001600_President_Trump_Speaks_at_National_Prayer_Breakfast.

35. Ibid.

36. Julian Borger, "'Brought to Jesus': The Evangelical Grip on the Trump Administration," *Guardian*, January 11, 2019, https://www.theguardian.com/us-news/2019/jan/11/trump-administration-evangelical-influence-support.

37. *The Godfather, Part III*, directed by Francis Ford Coppola, produced by Paramount Pictures and Zoetrope Studies, distributed by Paramount Pictures, December 25, 1990.

38. See: https://www.youtube.com/watch?v=yUVKRWXC5XM.

39. David Kertzer, *The Pope and Mussolini: The Secret History of Pius XI and the Rise of Fascism in Europe* (New York: Random House, 2014).

40. Benito Mussolini, in his essay of 1932, "Doctrine of Fascism," available online at http://www.historyguide.org/europe/duce.html.

41. Cited in George Orwell, *Orwell on Truth* (London: Harvill Secker, 2017), 26.

42. See the Sinclair Lewis Society, https://english.illinoisstate.edu/sinclairlewis/.

43. Ben Chapman, "How Trump Gets Away with Lying, as Explained by a Magician," *Medium*, June 27, 2018, https://medium.com/s/story/how-trump-gets-away-with-lying-as-explained-by-a-magician-4a14570fe6b0.

44. Machiavelli, *The Prince*.

CHAPTER 6

1. Niccolò Machiavelli, *The Prince*, trans. W. K. Marriott (originally 1513), chap. 10, available online at Project Gutenberg: http://www.gutenberg.org/files/1232/1232-h/1232-h.htm.

2. See the report by Sophie Tatum, "Trump Defends Putin," *CNN Politics*, February 6, 2017, https://www.cnn.com/2017/02/04/politics/donald-trump-vladimir-putin/index.html.

3. Rebecca Solnit, "The American Civil War Didn't End. And Trump Is a Confederate President," *Guardian*, November 4, 2018, https://www.theguardian.com/commentisfree/2018/nov/04/the-american-civil-war-didnt-end-and-trump-is-a-confederate-president.

4. Machiavelli, *The Prince*.

5. Ibid., chap. 16.

6. Benito Mussolini, in *Diuturna* (1921), quoted in H. B. Veatch, *Rational Man: A Modern Interpretation of Aristotelian Ethics* (Bloomington: Indiana University Press, 1962).

7. Orly Kayam, "The Readability and Simplicity of Donald Trump's Language," *Political Studies Review* 16 (2018): 2–12.

8. Basil Bernstein, *Class, Codes and Control: Theoretical Studies towards a Sociology of Language* (London: Routledge, 1971).

9. Elizabeth Hardwick, *Bartleby in Manhattan and Other Essays* (New York: Vintage, 1968), 46.

10. Donald J. Trump (@realDonaldTrump), Twitter, November 18, 2018, 1:01 p.m., https://twitter.com/realdonaldtrump/status/1064216956679716864?lang=en.

11. Allan Bloom, *Closing of the American Mind: How Higher Education Has Failed Democracy and Impoverished the Souls of Today's Students* (New York: Simon & Schuster, 1987).

12. Kat Chow, "'Politically Correct': The Phrase Has Gone from Wisdom to Weapon," NPR, *Code Switch*, December 14, 2016, https://www.npr.org/sections/codeswitch/2016/12/14/505324427/politically-correct-the-phrase-has-gone-from-wisdom-to-weapon.

13. George H. W. Bush, remarks at the University of Michigan Commencement Ceremony in Ann Arbor, May 4, 1991, George H. W. Bush Presidential Library.

14. Dinesh D'Souza, *Illiberal Education: The Politics of Race and Sex on Campus* (New York: Free Press, 1991).

15. Ruth King, *Talking Gender: A Guide to Nonsexist Communication* (Toronto: Copp Clark Pitman, 1991), 27.

16. Deborah Tannen, *Framing in Discourse* (Oxford: Oxford University Press, 1993), 77.

17. Donald J. Trump, reported by *The New York Times*, https://www.nytimes.com/live/republican-debate-election-2016-cleveland/trump-on-political-correctness/.

18. See the relevant report by Jessica Taylor, "Trump Calls for 'Total and Complete Shutdown of Muslims Entering' U.S.," *NPR*, December 7, 2015, https://www.npr.org/2015/12/07/458836388/trump-calls-for-total-and-complete-shutdown-of-muslims-entering-u-s.

19. *Adventures of Superman*, American Broadcasting Corporation television series, from 1952 to 1958.

20. Ruth Perry, "Historically Correct," *Women's Review of Books* 9 (1992): 15–16.

21. Ibid., 15.

22. Bloom, *Closing of the American Mind*, 33–34.

23. Antonio Gramsci, *Lettere dal carcere* (Torino: G. Einaudi, 1947).

24. Machiavelli, *The Prince*, chap. 5.

25. Ibid., chap. 6.

26. Niccolò Machiavelli, *Discourses on the First Decade of Titus Livius*, originally 1531, translated by Ninian Hill Thomson (London: Kegan Paul, Trench and Company, 1883), Book II, chapter 5.

27. Antonio Gramsci, *Lettere dal carcere, 1926–1937* (Torino: Einaudi, 1947).

28. Donald J. Trump (@realDonaldTrump), Twitter, December 10, 2018, 6:46 a.m., https://twitter.com/realdonaldtrump/status/1072095127894667265?lang=en.

29. Donald J. Trump (@realDonaldTrump), Twitter, August 14, 2018, 6:31 a.m., https://twitter.com/realdonaldtrump/status/1029329583672307712?lang=en.

30. Donald J. Trump (@realDonaldTrump), Twitter, October 17, 2012, 1:47 p.m., https://twitter.com/realdonaldtrump/status/258640349872926720?lang=en.

31. Donald J. Trump (@realDonaldTrump), Twitter, April 6, 2015, 10:22 p.m., https://twitter.com/realdonaldtrump/status/585281558816370690?lang=en.

32. Reported in *USA Today*, August 12, 2015, https://www.usatoday.com/story/news/politics/elections/2015/08/12/fact-check-trump-comments-women-megyn-kelly/31525419/.

33. See, respectively, Donald J. Trump (@realDonaldTrump), Twitter, February 28, 2016, 8:07 a.m. https://twitter.com/realdonaldtrump/status/703974892237135873?lang=en, and Twitter, November 18, 2018, 10:01 a.m., https://twitter.com/realdonaldtrump/status/1064216956679716864?lang=en.

34. *Time*, November 28, 1949, 16.

35. Christian R. Hoffman, "Crooked Hillary and Dumb Trump: The Strategic Use and Effect of Negative Evaluations in US Election Campaign Tweets," *Internet Pragmatics* 1 (2018): 55–87.

36. For example, Donald J. Trump (@realDonaldTrump), Twitter, October 1, 2017, 12:01 p.m., https://twitter.com/realdonaldtrump/status/914565910798782465?lang=en.

37. Elton John, "Rocket Man," released 1972, from the album, *Honky Château*, on the DJM label.

38. "Rocket Man," *The Economist*, July 6, 2006, https://www.economist.com/leaders/2006/07/06/rocket-man. https://twitter.com/realdonaldtrump/status/1014090584963866624?lang=en.

39. Sigmund Freud, *Civilization and Its Discontents* (London: Hogarth, 1963), 235–36.

40. For example, Donald J. Trump (@realDonaldTrump), Twitter, July 3, 2018, 3:16 a.m.

41. For instance, Donald J. Trump (@realDonaldTrump), Twitter, December 13, 2018, 9:34 a.m., https://twitter.com/realdonaldtrump/status/1073269881334849536?lang=en.

42. Donald J. Trump (@realDonaldTrump), Twitter, October 16, 2018, 8:04 a.m., https://twitter.com/realdonaldtrump/status/1052213711295930368?lang=en.

43. See *YouTube:* https://www.youtube.com/watch?v=xKevXvC_Eik.

44. Donald J. Trump (@realDonaldTrump), Twitter, November 9, 2018, 9:10 a.m., https://twitter.com/realdonaldtrump/status/1060942600994050048?lang=en.

45. Donald J. Trump (@realDonaldTrump), Twitter, January 15, 2018, 12:28 p.m., https://twitter.com/realdonaldtrump/status/953000902331453442?lang=en.

46. Donald J. Trump (@realDonaldTrump), Twitter, August 3, 2018, 8:37 p.m., https://twitter.com/realdonaldtrump/status/1025586524782559232?lang=en.

47. See Michael Collins and Christal Haye, "Here Are the Times Donald Trump's Critics Say He Stoked Racial Tensions," *USA Today*, August 5, 2019, https://www.usatoday.com/story/news/politics/2019/08/05/trump-and-race-presidents-critics-say-he-has-stoked-racial-tensions/1921410001/.

48. Chris Cillizza, "The Dangerous Consequences of Trump's All-Out Assault on Political Correctness," CNN, October 30, 2018, https://www.cnn.com/2018/10/30/politics/donald-trump-hate-speech-anti-semitism-steve-king-kevin-mccarthy/index.html.

49. Joe Palazzolo, Michael Rothfeld, and Lukas I. Alpert, "National Enquirer Shielded Donald Trump from Playboy Model's Affair Allegation," *Wall Street Journal*, November 4, 2016, https://www.wsj.com/articles/national-enquirer-shielded-donald-trump-from-playboy-models-affair-allegation-1478309380.

50. Jennifer Mercieca, "The Denials of Donald Trump," *Houston Chronicle*, December 11, 2018, https://www.houstonchronicle.com/local/gray-matters/article/donald-trump-robert-mueller-rhetorical-strategy-13458032.php.

51. Ibid.

52. Machiavelli, *The Prince*, chap. 18.

53. Ken Kesey, *One Flew over the Cuckoo's Nest* (New York: Signet, 1962).

54. D. Casarett, A. Pickard, J. M. Fishman, et al., "Can Metaphors and Analogies Improve Communication with Seriously Ill Patients?" *Journal of Palliative Medicine* 13 (2010): 255–60.

55. A. Byrne, J. Ellershaw, C. Holcombe, and P. Salmon, "Patients' Experience of Cancer: Evidence of the Role of 'Fighting' in Collusive Clinical Communication," *Patient Education and Counseling* 48 (2002): 15–21.

56. Linda Rogers, *Wish I Were: Felt Pathways of the Self* (Madison: Atwood, 1998).

57. For a general treatment of this topic, see Marcel Danesi and Nicolette Zukowski, *Medical Semiotics* (Munich: Lincom Europa, 2019).

58. George Lakoff, "The Contemporary Theory of Metaphor," in *Foundational Texts in Linguistic Anthropology*, ed. M. Danesi and S. Maida-Nicol (Toronto: Canadian Scholars, 2012), 163–64.

59. Gilles Fauconnier and Mark Turner, *The Way We Think: Conceptual Blending and the Mind's Hidden Complexities* (New York: Basic, 2002).

60. Susan Sontag, *Illness as Metaphor* (New York: Farrar, Straus & Giroux, 1978).

61. Michael Kranish, *Trump Revealed: An American Journey of Ambition, Ego, Money, and Power* (New York: Scribner, 2016).

62. Thomas F. Pettigrew, "Social Psychological Perspectives on Trump Supporters," *Journal of Social and Political Psychology* 5, no. 1 (2017), https://jspp.psychopen.eu/article/view/750/html; an insightful discussion of this very study is the one by Bobby Azarian, "An Analysis of Trump Supporters Has Identified 5 Key Traits: A New Report Sheds Light on the Psychological Basis for Trump's Support," *Psychology Today*, December 31, 2017, https://www.psychologytoday.com/ca/blog/mind-in-the-machine/201712/analysis-trump-supporters-has-identified-5-key-traits.

63. See Cillizza, "The Dangerous Consequences of Trump's All-Out Assault on Political Correctness," op. cit.

64. *Don't Be a Sucker*, distributed by the United States Department of War, July 4, 1943, https://www.youtube.com/watch?v=vGAqYNFQdZ4.

65. Machiavelli, *The Prince*, chap. 18.

66. Thomas Hobbes, *Elements of Philosophy* (London: Molesworth, 1656).

CHAPTER 7

1. Donald J. Trump and Tony Schwartz, *The Art of the Deal* (New York: Ballantine, 1987), chap. 2.

2. Claudia Claridge, *Hyperbole in English: A Corpus-Based Study of Exaggeration* (Cambridge: Cambridge University Press, 2011).

3. Marty Neumeier, *The Brand Gap* (Berkeley: New Riders, 2006), 39.

4. See YouTube: https://www.youtube.com/watch?v=spKkzjD1SEU.

5. See YouTube: https://www.youtube.com/watch?v=olh3GdUUk24.

6. See transcript of the first presidential debate, https://www.latimes.com/politics/la-na-pol-debate-transcript-clinton-trump-20160926-snap-htmlstory.html.

7. See: https://en.wikiquote.org/wiki/Donald_Trump.

8. P. T. Barnum, *Struggles and Triumphs* (Hartford: J. B. Burr, 1869), 72.

9. S. Romi Mukherjee, "Make America Great Again as White Political Theology," *LISA E-Journal* 16, no. 2 (2018), https://journals.openedition.org/lisa/9887.

10. Michael Wolff, *Fire and Fury: Inside the Trump White House* (New York: Henry Holt, 2018).

11. Chauncey DeVega, "Why Do Evangelicals Worship Trump? The Answer Should Be Obvious," AlterNet, January 13, 2018, https://www.alternet.org/news-amp-politics/why-do-evangelicals-worship-trump-answer-should-be-obvious.

12. Elisabeth R. Anker, *Orgies of Feeling: Melodrama and the Politics of Feeling* (Durham: Duke University Press, 2014).

13. Reported by Jack Jenkins, "Trump's 'God Whisperer' Says Resisting Him Is an Affront to God," Thinkprogress.org, August 23, 2017, https://thinkprogress.org/trump-spiritual-adviser-affront-god-d615c512bffc/.

14. Reza Aslan, "The Dangerous Cult of Donald Trump," *Los Angeles Times*, November 6, 2017, https://www.latimes.com/opinion/op-ed/la-oe-aslan-trump-cultists-20171106-story.html.

15. P. T. Barnum, *The Life of P. T. Barnum* (New York: Redfield, 1854).

16. Kevin Young, *Bunk: The Rise of Hoaxes, Humbug, Plagiarists, Phonies, Post-Facts, and Fake News* (Minneapolis: Graywolf, 2017).

17. James W. Pennebaker, *The Secret Life of Pronouns* (London: Bloomsbury, 2011).

18. See the report by Miriam Valverde, "Donald Trump Says Chicago Has Experienced Thousands of Shootings This Year," *Politifact*, September 27, 2016, https://www.politifact.com/truth-o-meter/statements/2016/sep/27/donald-trump/donald-trump-says-chicago-has-experienced-thousand/.

19. See transcript of the second debate: https://www.nytimes.com/2016/10/10/us/politics/transcript-second-debate.html.

20. Reported by *ReverbPress*, September 27, 2016, https://reverbpress.com/politics/heres-a-little-sunshine-on-trumps-solar-nonsense-video/.

21. From the ABC *This Week* interview, July 31, 2016, https://abcnews.go.com/Politics/week-transcript-donald-trump-vice-president-joe-biden/story?id=41020870.

22. Dell Hymes, *On Communicative Competence* (Philadelphia: University of Pennsylvania Press, 1971).

23. Theodore Millon, *Disorders of Personality*, 3rd ed. (Hoboken, NJ: Wiley, 2011).

24. These are discussed in more detail in Michael Arntfield and Marcel Danesi, *Murder in Plain English: From Manifestos to Memes—Looking at Murder through the Words of Killers* (New York: Prometheus, 2017).

25. See YouTube: https://www.cnn.com/videos/us/2018/03/23/karen-mcdougal-full-interview-ac.cnn.s.

26. See, for example, "Donald Trump and the Self-Made Man," *The New York Times*, October 2, 2018, https://www.nytimes.com/2018/10/02/opinion/donald-trump-tax-fraud-fred.html.

27. Phone interview on "Fox and Friends," quoted in Jesse Byrnes, "Trump on Obama and Islam: 'There's Something Going In,'" *The Hill*, June 13, 2016, https://thehill.com/blogs/blog-briefing-room/news/283246-trump-on-obama-and-islam-theres-something-going-on.

28. Joseph Burgo, "The Populist Appeal of Trump's Narcissism: Why Trump's Narcissistic Personality Attracts Disaffected Voters," *Psychology Today*, August 14, 2015, https://www.psychologytoday.com/ca/blog/shame/201508/the-populist-appeal-trumps-narcissism.

29. Herman Melville, *The Confidence-Man: His Masquerade* (London: Longman, 1857).

30. George Ade, *Artie* (Chicago: Herbert S. Stone & Co., 1896).

31. Max Boot, "Trump Spent His Business Career Swindling People. Nothing's Changed," *Chicago Tribune*, May 3, 2018, https://www.chicagotribune.com/news/opinion/commentary/ct-donald-trump-business-scams-20180503-story.html.

32. Max Boot, "Trump Is a Grifter—Same as Ever," *Washington Post*, May 2, 2018, https://www.washingtonpost.com › opinions › global-opinions.

33. Hannah Arendt, *The Human Condition* (Chicago: University of Chicago Press, 1958).

34. Walter Lippmann, *Public Opinion* (New York: Macmillan, 1922).

35. Charles Dickens, *Oliver Twist* (London: Richard Bentley, 1837).

36. Niccolò Machiavelli, *The Prince*, trans. W. K. Marriott (originally 1513), chap. 18, available online at Project Gutenberg, http://www.gutenberg.org/files/1232/1232-h/1232-h.htm.

37. Ben Zimmer, "Donald Trump and the Art of the 'Con,'" *Atlantic*, September 27, 2018, https://www.theatlantic.com/ideas/archive/2018/09/donald-trump-and-the-art-of-the-con/571528/.

38. David Maurer, *The Big Con: The Story of the Confidence Man* (New York: Anchor, 1940).

39. *The Sting*, directed by George Roy Hill, produced by Universal Pictures, released December 25, 1973.

40. *The Hucksters*, directed by Jack Conway, produced by Metro-Goldwyn-Mayer, released August 27, 1947.

41. *A Letter to Three Wives*, directed by Joseph L. Mankiewicz, 1949, produced by Twentieth Century Fox, released January 20, 1949.

42. *A Face in the Crowd*, directed by Elia Kazan, produced by Warner Brothers, released May 28, 1957.

43. Max Black, "The Prevalence of Humbug," *Philosophic Exchange* 13, no. 1 (1982), https://digitalcommons.brockport.edu/phil_ex/vol13/iss1/4.

44. T. S. Eliot, "The Triumph of Bullshit," http://poetry-fromthehart.blogspot.com/2011/08/triumph-of-bullshit-ts-eliot.html.

45. Quoted in German Lopez, "Trump Keeps Lying about His Electoral College Victory," *Vox*, February 16, 2017, https://www.vox.com/policy-and-politics/2017/2/16/14639058/trump-electoral-college-win.

46. Reported by Philip Bump in the *Washington Post*, February 6, 2017, https://www.washingtonpost.com > news > politics.

47. James Ball, *Post-Truth: How Bullshit Conquered the World* (London: Biteback, 2017).

48. Harry G. Frankfurt, *On Bullshit* (Princeton: Princeton University Press, 2005), xiii.

49. Black, "Prevalence of Humbug."

50. Stanton Peele, "Bullshitting: Lessons from the Masters," *Psychology Today*, May 15, 2009, https://www.psychologytoday.com/ca/blog/addiction-in-society/200905/bullshitting-lessons-the-masters.

51. Jason Pine, *The Art of Making Do in Naples* (Minneapolis: University of Minnesota Press, 2012).

52. Michel de Montaigne, *Essays*, translated by Charles Cotton, originally 1580 (London: Reeves and Turner, 1877).

53. Dallas G. Denery, *The Devil Wins: A History of Lying from the Garden of Eden to the Enlightenment* (Princeton: Princeton University Press, 2015).

54. Jean-Jacques Rousseau, *Discourse on the Origins of Inequality* (London: R. and J. Dodsley, 1581).

55. Robert Louis Stevenson, *Virginibus Puerisque and Other Papers* (London: C. Kegan Paul, 1881), chap. 2.

56. Vance Packard, *The Hidden Persuaders* (New York: McKay, 1957).

57. J. B. Twitchell, *20 Ads That Shook the World* (New York: Crown, 2000), 23.

58. Stuart Ewen, *All Consuming Images* (New York: Basic, 1988), 20.

59. Bob Stein, "We Could Be Better Ancestors Than This: Ethics and First Principles for the Art of the Digital Age," in *The Digital Dialectic: New Essays on New Media*, ed. Peter Lunenfeld (Cambridge: MIT Press, 1999), 204.

60. See YouTube: https://www.youtube.com/watch?v=a5pYa5cxLEo.

61. Sarah Churchwell, *Behold America: The Entangled History of "America First" and "The American Dream"* (London: Bloomsbury, 2018).

62. Timothy O'Brien, *TrumpNation: The Art of Being the Donald* (New York: Grand Central, 2005).

CHAPTER 8

1. Frans de Waal, *Chimpanzee Politics* (Baltimore: Johns Hopkins University Press, 1982).

2. Max Weber, *Theory of Social and Economic Organization* (New York: Free Press, 1922), 328.

3. Cited in George Orwell, *Orwell on Truth* (London: Harvill Secker, 2017), 75.

4. Niccolò Machiavelli, *The Prince*, trans. W. K. Marriott (originally 1513), chap. 18, available online at Project Gutenberg: http://www.gutenberg.org/files/1232/1232-h/1232-h.htm.

5. Innocent Gentillet, *Discours contre Machievel*, originally 1576, translated by Simon Patericke (Eugene, OR: Resource Publications, 2018).

6. William Shakespeare, *Henry VI, Part III*, Act 3, Scene 2.

7. Denis Diderot, "Machiavelisme," in *Encyclopédie ou Dictionnaire raisonné des sciences, des arts et des métiers*, vol. 9 (Paris, 1765).

8. Richard Christie and Florence L. Geis, *Studies in Machiavellianism* (New York: Academic Press, 1970).

9. See Harley Therapy, "What is Machiavellianism in Psychology?" *Counseling Blog*, January 8, 2015, https://www.harleytherapy.co.uk/counselling/machiavellianism-psychology.htm.

10. Peter K. Jonason and James P. Middleton, "Dark Triad: The 'Dark Side' of Human Personality," in *International Encyclopedia of the Social & Behavioral Sciences*, 2nd ed. (Oxford: Elsevier, 2015), 671–75.

11. Glenn Geher, "Donald Trump as High in the Dark Triad," *Psychology Today*, August 6, 2016, https://www.psychologytoday.com/ca/blog/darwins-subterranean-world/201608/donald-trump-high-in-the-dark-triad.

12. A. Vartanian, *La Mettrie's "L'homme machine": A Study in the Origins of an Idea* (Princeton: Princeton University Press, 1960).

13. Jaclyn Duffin, *Lovers and Livers: Disease Concepts in History* (Toronto: University of Toronto Press, 2005).

14. Anita Kelly and Lijuan Wang, "A Life without Lies: How Living Honestly Can Affect Health" (presentation at the Annual American Psychological Association Convention, Orlando, FL, session 3189, August 2–5, 2012).

15. Gregory Bateson, *Steps to an Ecology of Mind* (New York: Ballantine, 1972).

16. Machiavelli, *The Prince.*

17. L. Carmichael, H. P. Hogan, and A. A. Walter, "An Experimental Study of the Effect of Language on Visually Perceived Form," *Journal of Experimental Psychology* 15 (1932): 73–86.

18. Ann Gill, *Rhetoric and Human Understanding* (Prospect Heights, IL: Waveland, 1994), 106.

19. Parmenides, "The Way of Truth," in John Burnet, *Early Greek Philosophy*, chapter 4 (London: Adam & Charles Black, 1892).

20. Leon Festinger, *A Theory of Cognitive Dissonance* (Evanston, IL: Row, Peterson, 1957).

21. Leon Festinger, Henry W. Riecken, and Stanley Schachter, *When Prophecy Fails* (London: Printer & Martin, 1956), 3.

22. Ibid., 3.

23. Benito Mussolini, quoted in Israel W. Charny, *Fascism and Democracy in the Human Mind: A Bridge Between Mind and Society* (Lincoln: University of Nebraska Press, 2006), 23.

24. Marshall McLuhan, spoken on the *Tomorrow Show* with Tom Snyder, originally aired on September 6, 1976, on NBC.

25. Neil Postman, *Technopoly: The Surrender of Culture to Technology* (New York: Alfred A. Knopf, 1992).

26. Jaron Lanier, *You Are Not a Gadget* (New York: Vintage, 2010), 12.

27. *The Matrix,* directed by the Wachowski Brothers, produced and released by Warner Brothers on March 31, 1999.

28. Manuel Castells, *The Information Age: Economy, Society, and Culture* (Oxford: Blackwell, 1996).

29. See Marshall McLuhan, *Understanding Media: The Extensions of Man* (Cambridge: MIT Press).

30. See Brooke Seipel, "Trump: 'Make America Great Again' Slogan 'Was Made Up by Me,'" *The Hill*, April 2, 2019, https://thehill.com/homenews/administration/437070-trump-make-america-great-again-slogan-was-made-up-by-me.

31. Michelle Tolman Clarke, "On the Woman Question in Machiavelli," *Review of Politics* 67 (2005): 229–55.

32. Florence Nightingale, *Suggestions for Thought to Searchers after Religious Truth*, ed. Michael D. Calabria and Janet A. McCrae (1852; repr., Philadelphia: University of Pennsylvania Press, 1994).

33. Edward O. Wilson, *On Human Nature* (New York: Bantam, 1979).

34. Michael Arntfield and Marcel Danesi, *Murder in Plain English: From Manifestos to Memes—Looking at Murder through the Words of Killers* (New York: Prometheus, 2017).

35. James Piereson, "A Note on Character in Politics," *Dispatch*, January 4, 2019, https://www.newcriterion.com/blogs/dispatch/a-note-on-character-in-politics.

36. Lewis Carroll, *Through the Looking Glass and What Alice Found There* (London: Macmillan, 1871).

37. Giambattista Vico, *The New Science of Giambattista Vico*, trans. and ed. Thomas G. Bergin and Max Fisch (1744; repr., Ithaca: Cornell University Press, 1984).

38. Plato, *The Republic of Plato*, originally 381 BCE, translated by J. L. Davies and D. J. Vaughan (London: Macmillan, 1888).

39. Humberto Maturana and Francisco Varela, *Autopoiesis and Cognition: The Realization of the Living* (Dordrecht: Reidel, 1973).

40. Lord Byron, *Childe Harold's Pilgrimage* (1812; repr., Boston: Houghton Mifflin, 1894), canto 2, stanza 2.

41. Milan Kundera, *Immortality* (London: Faber and Faber, 1991), 239.

42. Hans Christian Andersen, *The Emperor's New Clothes* (Denmark: C. A. Reitzel, 1837).

43. Orwell, *Orwell on Truth*, 129.